CELLULAR PHYSIOLOGY
OF NERVE & MUSCLE

CELLULAR PHYSIOLOGY OF NERVE & MUSCLE

Gary G. Matthews

Department of Neurobiology and Behavior
State University of New York
at Stony Brook

Blackwell Scientific Publications
Palo Alto   Oxford   London   Edinburgh   Boston   Victoria

For Karen and David

Editorial Offices

667 Lytton Avenue
Palo Alto, California 94301, USA

Osney Mead, Oxford, OX2 0EL, UK

8 John Street
London, WC1N 2ES, UK

23 Ainslie Place
Edinburgh, EH3 6AJ, UK

52 Beacon Street
Boston, Massachusetts 02108, USA

107 Barry Street
Carlton, Victoria 3053, Australia

Distributors

USA and Canada
Blackwell Scientific Publications
P.O. Box 50009
Palo Alto, California 94303

Australia
Blackwell Scientific Publications
(Australia) Pty Ltd.
107 Barry Street
Carlton, Victoria 3053

United Kingdom
Blackwell Scientific Publications
Osney Mead
Oxford OX2 0EL

Sponsoring Editor: John Staples
Manuscript Editor: Sean Cotter
Production Coordinator: Robin Mitchell
Artist: Georg Klatt
Interior and Cover Design: Gary Head
Composition: Jonathan Peck Typographers, Ltd.
Printing and Binding: Fairfield Graphics

First Published 1986.

© 1986 by Blackwell Scientific Publications

Library of Congress Cataloging in Publication Data

Matthews, Gary G., 1949–
  Cellular physiology of nerve & muscle.

  Bibliography: p.
    1. Nerves—Cytology. 2. Muscles—Cytology.
3. Electrophysiology 4. Neural conduction 5. Cell
physiology. I. Title. II. Title: Cellular
physiology of nerve and muscle. [DNLM: 1. Cell
Membrane—physiology. 2. Membrane Potentials.
3. Muscles—physiology. 4. Neural Transmission.
5. Neurons—physiology. QH 601 M439c]
QP363.M38 1985   599′.018   85-15727
ISBN 0-86542-309-1

Preface

The focus of the material in this book is the electrical behavior of nerve and muscle cells. It is intended to introduce the subject to students who have completed an introductory biology course and who understand a few basic facts about chemistry and physics. It is *not* intended, however, to be a comprehensive survey of nerve and muscle physiology; rather, I have tried to limit the material to the basic physical and chemical principles underlying the functioning of the nervous system and of muscles. In my experience as a teacher, the origin of the resting membrane potential and the electrical mechanisms of the action potential and of synaptic transmission are among the most difficult topics for typical biology students to master. Part of the problem is that beginners have difficulty extracting the basic principles from the array of facts and experimental results inherent in a comprehensive presentation of nerve–muscle physiology. My approach is to develop the principles necessary for an understanding of this material through a sequence of model systems and schematic distillations of experimental results. I have included descriptions of actual experiments where I think they might be interesting or provide a good example to illustrate a principle, but for the most part I construct hypothetical experiments tailored to elucidate the point at hand. Similarly, I have not shied from quantitative material when it is necessary and can be integrated into a developing presentation of a topic, but I have not included equations for their own sake.

I hope this book will provide a framework around which students can organize the many facts about the nervous system, muscles, and their interactions that must be mastered for a thorough understanding of nervous system function.

My lectures on electrical properties of cells, and hence the organization of this book, have been greatly influenced by my own teachers: Denis Baylor and John Nicholls at Stanford; William Betz, Bob Martin, and Warren Wickelgren at the University of Colorado; and Randy Gallistel at the University of Pennsylvania. Their teaching of difficult material set standards of clarity and organization that I hope are reflected in this book. Any failure to meet those standards is, of course, my own doing. Special thanks are due the numerous reviewers who carefully read earlier versions of the manuscript and gave much valuable advice.

Gary Matthews

Contents

PART III
CELLULAR PHYSIOLOGY OF
MUSCLE CELLS 147

CELLULAR PHYSIOLOGY
OF NERVE & MUSCLE

PART I

ORIGIN OF ELECTRICAL MEMBRANE POTENTIAL

This is a book about some of the physiological characteristics of nerve and muscle cells. As we shall see, many of the important functions of these cells are carried out by electrical mechanisms. That is, the properties of these cells as generators and conductors of electricity are vital to their functioning. Our goal, then, is to arrive at an understanding of the basic physical and chemical principles underlying the electrical behavior of nerve and muscle cells and to apply those principles to the normal physiological functioning of those cells in the body.

Because an understanding of how electrical voltages and currents arise in cells is central to this book, Part I is devoted to that task. The discussion begins with a presentation of the differences in composition between fluids inside and outside cells and then considers the factors that are important in achieving osmotic and electrochemical equilibrium states. We will then see that animal cells are not at equilibrium, but rather must expend energy in order to maintain the differing compositions of their internal and external fluids. By the end of Part I we will arrive at a quantitative description of the relation between ionic gradients and electrical gradients across cell membranes. This quantitative description sets the stage for the specific descriptions of nerve and muscle cells in the remaining chapters of the book and is central to understanding how the nervous system functions as a transmitter of electrical signals.

Part II examines cellular properties of individual nerve cells and their interactions, again emphasizing the electrical properties developed in Part I. Part III turns to muscle cells and their control by the nervous system. In this case, however, the emphasis is on the linkage between the electrical signal in a single muscle and the mechanical contraction and on the mechanical properties that are important in the regulation of contraction by the nervous system. Part III also describes the electrical differences between the muscle cells of the heart and those of other muscles and relates those differences to the specialized function of the heart muscle.

Throughout the book, little is assumed in terms of background knowledge in physics, chemistry, or mathematics. Important quantitative principles are derived from examples or explained in a qualitative as well as quantitative manner. It is assumed, however, that the reader has completed an introductory, college-level course in biology and understands basic characteristics of cells and cellular metabolism.

1

Composition of Intracellular and Extracellular Fluids

When we think of biological molecules, we normally think of all the special molecules that are unique to living organisms, such as proteins and nucleic acids: enzymes, DNA, RNA, and so on. These are the substances that allow life to occur and give living things their special characteristics. Yet, if we were to dissociate a human body into its component molecules and sort them by type, we would find that these special molecules are only a small minority of the total. Of all the molecules in a human body, only about 0.25% fall within the category of these special biological molecules. Most of the molecules are far more ordinary. In fact, the most common molecule in the body is water. Excluding nonessential body fat, water makes up about 75% of the weight of a human body. Because water is a comparatively light molecule, especially when compared with massive protein molecules, this 75% of body weight translates into a staggering number of molecules of water. Thus, water accounts for about 99% of all molecules in the body. The remaining 0.75% consists of other simple inorganic substances, mostly sodium, potassium, and chloride ions. In the first part of this book, we will be concerned in large part with the mundane majority of molecules, the 99.75% made up of water and inorganic ions.

Why should we study these mundane molecules? Although many enzymatic reactions involving the more glamorous organic molecules require the participation of inorganic cofactors, and although most biochemical reactions within cells occur among substances that are dissolved in water, most inorganic molecules in the body never participate in any biochemical reactions. In spite of this, a sufficient reason to study these substances is that cells could not exist and life as we know it would not be possible if cells did not possess mechanisms to control the distribution of water and ions across their membranes. The purpose of the first four chapters of this book is to see why that is true and to understand the physical principles that underlie the ability of cells to maintain their integrity in a hostile physicochemical environment.

Intracellular and Extracellular Fluids

The water in the body can be divided into two compartments: intracellular and extracellular fluid. About 55% of the water is inside cells, and the remainder is outside. The extracellular fluid, or ECF, can in turn be subdivided into plasma, lymphatic fluid, and interstitial fluid, but for now we can lump all the ECF together into one compartment. Similarly, there are subcompartments within cells, but it will suffice for now to treat cells as uniform bags of fluid. The wall that separates the intracellular and extracellular fluid compartments is the outer cell membrane, also called the **plasma membrane** of the cell.

There are both organic and inorganic substances dissolved in the intracellular and extracellular water, but the compositions of the two fluid compartments are different. Table 1-1 gives simplified compositions of ECF and intracellular fluid (ICF) for a typical mammalian cell. The table is simplified because it includes only those substances that are important in governing the basic osmotic and electrical properties of cells. There are many more kinds of inorganic and organic solutes in both ECF and ICF, and many of them have important physiological roles in other contexts. For the present, however, they can be safely ignored.

The principal cation (positively charged ion) outside the cell is sodium, although there is a small amount of potassium, which will be important to consider when we discuss the origin of the membrane potential of cells. Inside cells, the situation is very different. In ICF, the principal cation is potassium, although there is a small amount of sodium. Negatively charged chloride ions, which are present at high concentration in ECF, are relatively scarce in ICF. The major anion (negatively charged ion) inside cells is actually a class of molecules that bear a net negative charge. These intracellular anions, which we will abbreviate A^-, include protein molecules, acidic amino acids like aspartate and glutamate, and inorganic ions like sulfate and phosphate. For the purposes of this

Table 1-1 Simplified compositions of intracellular and extracellular fluids for a typical mammalian cell.

	Internal concentration (m*M*)	External concentration (m*M*)	Can it cross plasma membrane?
K^+	125	5	Y
Na^+	12	120	N*
Cl^-	5	125	Y
A^-	108	0	N
H_2O	55,000	55,000	Y

Note: Membrane potential = -60 to -100 mV
*As we will see in Chapter 3, this No is not as simple as it first appears.

book, the anions of this class outside cells can be ignored; we will say that the only extracellular anion is chloride.

It will also be important to consider the concentration of water on the two sides of the membrane, which is also shown in Table 1-1. It may seem odd to speak of the "concentration" of the solvent in ECF and ICF. However, as we shall see when we consider the maintenance of cell volume, the concentration of water inside and outside the cell must be the same or water will move across the membrane and cell volume will change.

Another important consideration will be whether or not a particular substance can cross the plasma membrane—that is, whether the membrane is permeable to that substance. The plasma membrane is permeable to water, potassium, and chloride, but is effectively impermeable to sodium (however, more will be said about sodium permeability later). Of course, if the membrane is to do its job properly, it must keep the organic anions inside the cell; otherwise all of a cell's essential biochemical machinery would simply diffuse away into the ECF. Thus, the membrane is impermeable to A^-

An ultrafine probe can be inserted inside an animal cell to measure the electrical voltage difference between the inside and outside of the cell. The inside is more negative than the outside. The difference is usually about 60 to 100 millivolts (mV), and is referred to as the **membrane potential** of the cell. By convention, the potential outside the cell is called zero; therefore, the typical value of the membrane potential (abbreviated E_m) is -60 to -100 mV, as shown in Table 1-1. A major concern of the first section of this book will be the origin of this electrical membrane potential. In later sections, we will discuss how the membrane potential influences the movement of charged particles across the cell membrane and how the electrical energy stored in the membrane potential can be tapped to generate signals that can be passed from one cell to another in the nervous system.

The Structure of the Plasma Membrane

Before we consider the mechanisms that allow cells to maintain the differences in ECF and ICF shown in Table 1-1, it will be helpful to look at the structure of the outer membrane of the cell, the plasma membrane. The control mechanisms responsible for the differences between ICF and ECF reside within that barrier between the intracellular and extracellular compartments.

It has long been known that the contents of a cell will leak out if the cell is damaged by being poked or prodded with a glass probe. Also, some dyes will not enter cells when dissolved in the extracellular fluid, and the same dyes will not leak out when injected inside cells. These observations, first made in the nineteenth century, led to the idea that there is a selectively permeable barrier—the plasma membrane—separating the intracellular and extracellular fluids.

The first systematic observations of the kinds of molecules that would enter cells and the kinds that were excluded were made by Overton in the early part of the twentieth century. He found that, in general, substances that are highly soluble in lipids enter cells more easily than substances that are less soluble in lipids. Lipids are molecules that are not soluble in water or other polar solvents, but are soluble in oil or other nonpolar solvents. Thus, Overton suggested that the plasma membrane of a cell is made of lipids and that substances can cross the membrane if they can dissolve in that lipid barrier.

There were some exceptions to the general lipid solubility rule. Electrically charged substances, like potassium and chloride ions, are almost totally insoluble in lipids, yet they manage to cross the plasma membrane. Other substances, such as urea, entered cells more easily than expected from their lipid solubility alone. To take account of these exceptions, Overton suggested that the lipid membrane is shot through with tiny holes or pores that allow highly water-soluble (hydrophilic) substances, such as ions, to cross the membrane. Only hydrophilic substances that are small enough to fit through these small aqueous pores can cross the membrane. Larger molecules like proteins and amino acids cannot fit through the pores and thus cannot cross the membrane without the help of special transport mechanisms.

The molecules of the lipid skin of cell membranes appear to be arranged in a layer only two molecules thick. Evidence for this arrangement was obtained from experiments in which the lipids were chemically extracted from the plasma membranes of cells and spread out on a trough of water in such a way that they formed a film only one molecule thick. When the area of this monolayer "oil slick" was measured, it was found to be about twice the total surface area of the intact cells from which the lipids were obtained. This suggests that the membrane of the intact cells was two molecules thick. Such a membrane is called a lipid bilayer membrane.

The bilayer arrangement of the cell membrane makes chemical sense when we consider the characteristics of the kinds of lipid molecules found in the plasma membrane. The cell lipids are largely phospholipids, which are molecules that have both a polar region that is hydrophilic and a nonpolar region that is hydrophobic. When surrounded by water, these lipid molecules tend to aggregate, with the hydrophilic regions oriented out toward the surrounding water and the hydrophobic regions pointed toward each other. When spread out in a sheet with water on each side of the sheet, the phospholipids can maintain their preferred state by forming a bimolecular sandwich, with the hydrophilic parts on the outside toward the water, and the hydrophobic parts in the middle, pointed toward each other. This bilayer model for the cell plasma membrane is illustrated in Figure 1-1.

Figure 1-1 also shows another important characteristic of cell membranes. They contain not only lipid molecules but also protein molecules. Some proteins are attached to the inner or outer surface of the cell membrane, and others penetrate all the way through the membrane so that they form a bridge from one side to the other. Some of the transmembrane proteins form the aqueous pores, or channels, that allow ions and other small hydrophilic molecules to cross the membrane. If we separate membranes from the rest of the cell and analyze their composition, we find that, by weight, only about one-third of the membrane material is lipid; most of the rest is protein. Thus, the lipids form the backbone of the membrane, but proteins are an important part of the picture. We will see later that the proteins are very important in controlling the movement of substances, particularly ions, across the cell membrane.

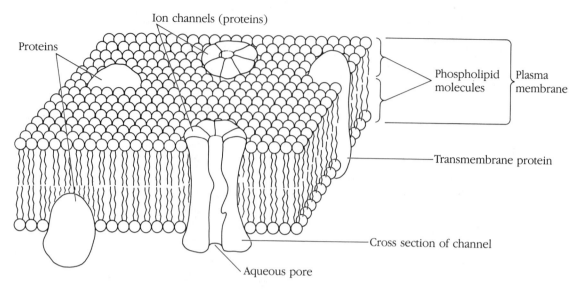

Proteins

Ion channels (proteins)

Phospholipid molecules

Plasma membrane

Transmembrane protein

Cross section of channel

Aqueous pore

Figure 1-1

Schematic diagram of a section of the plasma membrane. The backbone of the membrane is a sheet of lipid molecules two molecules thick. Inserted into this sheet are various types of protein molecules. Some protein molecules extend all the way across the sheet from the inner to the outer face. These transmembrane proteins sometimes form small-diameter aqueous pores or channels through which small hydrophilic molecules, such as ions, can cross the membrane. In the diagram, two such channels are shown, one cut in cross section to reveal the interior of the pore.

Anatomical evidence also supports the model shown in Figure 1-1. The cell membrane is much too thin to be seen with the light microscope. In fact, it is almost too thin to be seen with the electron microscope. However, with an electron microscope it is possible to see at the outer boundary of a cell a three-layered (trilaminar) profile like a railroad track, with a light region separating two darker bands. Figure 1-2 is an example of an electron micrograph showing the plasma membranes of two cells lying in close contact. The interpretation of the trilaminar profile is that the two dark bands represent the polar heads of the membrane phospholipids and protein molecules on the inner and outer surfaces of the membrane and that the lighter region between the two dark bands represents the nonpolar tails of the lipid molecules. The total thickness of the sandwich is about 75 Å. The lighter-colored "fuzz" surrounding the trilaminar profiles of the two cell membranes in Figure 1-2 consists in part of portions of membrane-associated protein molecules extending out into the intracellular and extra-cellular spaces. The two cells shown in Figure 1-2 are nerve cells (neurons) in the brain, and the region of close contact is a special-ized junction, called a synapse, where electrical activity is relayed from one nerve cell to another. One of our major goals in this book is to understand how such synapses, the basic mechanism of infor-mation transfer in the brain, work.

By using a special form of microscopy called freeze-fracture electron microscopy, it is possible to visualize more clearly the protein molecules that are imbedded in the plasma membrane. A

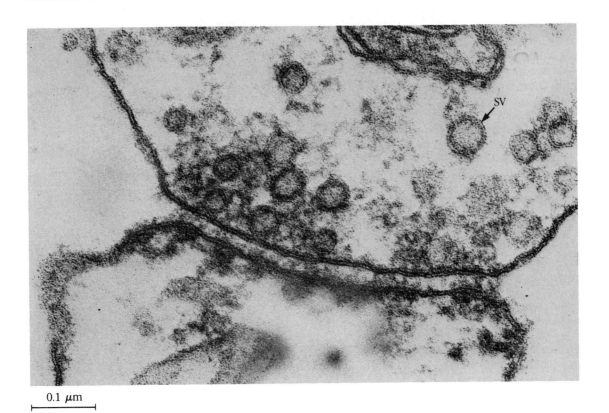

0.1 μm

Figure 1-2
High-power electron micrograph of the plasma membranes of two neighboring cells. Note the two dark bands separated by a light region at the outer surface of each cell. The two cells are nerve cells from the brain, and the point of close contact between them is a **synapse**, the point of information transfer in the nervous system. Note also the membrane-bound intracellular structures (labeled SV), called **synaptic vesicles**, inside one of the cells; the vesicle membranes also have the trilaminar profile seen in the plasma membranes. We will learn more about synaptic vesicles and synapses in Chapters 7 and 8. [Courtesy of A. de Blas of the State University of New York, Stony Brook.]

schematic representation of the freeze-fracture technique is shown in Figure 1-3. A small sample of the tissue to be examined is frozen in liquid nitrogen, and then a thin sliver of the frozen tissue is shaved off with a sharp knife. Because the tissue is frozen, however, the sliver is not so much sliced off as broken off from the sample. In some cases, like that shown in Figure 1-3, the line of fracture will run between the two lipid layers of the membrane bilayer, leaving holes where protein molecules are ripped out of the lipid monolayer and protrusions where membrane proteins are ripped out of the opposing monolayer and come along with the shaved sliver. An example of such a freeze-fracture sample viewed through the electron microscope is shown in Figure 1-4. The membrane proteins appear as small bumps in the otherwise smooth surface of the plasma membrane, like grains of sand sprinkled on a freshly painted surface. In the discussion of the transmission of signals from one nerve cell to another in Chapter 7, we will see other examples of freeze-fracture electron micrographs and see how they can provide important evidence about the physiological functioning of cells.

Figure 1-3
Schematic illustration of the freeze-fracture procedure for electron microscopy. When a fracture line runs between the two lipid layers of the plasma membrane, some membrane proteins stay with one monolayer, others with the other layer. If the fractured surface is then examined with the electron microscope, the remaining proteins appear as protruding bumps in the surface.

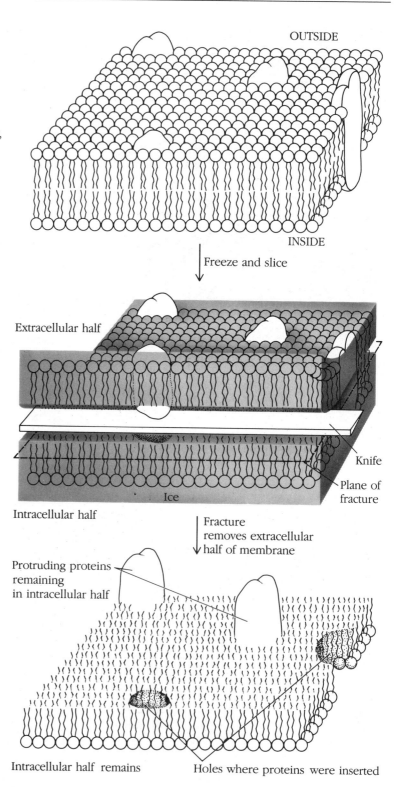

OUTSIDE

INSIDE

Freeze and slice

Extracellular half

Knife

Plane of fracture

Intracellular half

Ice

Fracture removes extracellular half of membrane

Protruding proteins remaining in intracellular half

Intracellular half remains

Holes where proteins were inserted

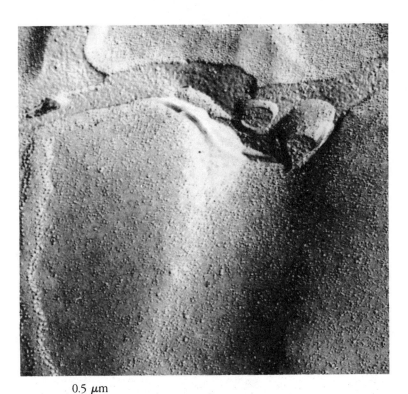

0.5 μm

Figure 1-4
Example of a fractured membrane surface containing protein molecules, viewed through the electron microscope. The membrane surface shown is that of the presynaptic nerve terminal at the nerve–muscle junction, which will be discussed in detail in Chapter 7. [Reproduced, with permission, from C.-P. Ko, *J. Cell Biol.* 98(1984):1685–1695.]

Summary

The most common molecules in the body are water and simple inorganic molecules—mainly sodium, potassium, and chloride ions. The water in the body can be divided into two compartments: the intracellular and extracellular fluids. The barrier between those two compartments is the plasma membrane of the cell, which is a phospholipid bilayer with protein molecules inserted into it. The extracellular fluid is high in sodium and chloride, but low in potassium, while the intracellular fluid is low in sodium and chloride, but high in potassium. This difference is maintained and regulated by control mechanisms residing in the plasma membrane, which act as a selectively permeable barrier permitting some substances to cross but excluding others.

2

Maintenance of Cell Volume

At an early stage of evolution, before the development of cells, life might well have been nothing more than a loose confederation of enzyme systems and self-replicating molecules. A major problem faced by such acellular systems must have been how to keep their constituent parts from simply diffusing away into the surrounding murk. The solution to this problem was the development of a cell membrane that was impermeable to these organic molecules; this was the origin of cellular life. However, the cell membrane, while solving one problem, brought with it a new problem: how to achieve osmotic balance. To see how this problem arises, it will be useful to begin with a review of solutions, osmolarity, and osmosis. We will then turn to an analysis of the cellular mechanisms used to deal with problems of osmotic balance.

Molarity, Molality, and Diffusion of Water

Examine the situation illustrated in Figure 2-1. We take one liter of pure water and dissolve some sugar in it. The dissolved sugar molecules take up some space that was formerly occupied by water molecules, and thus the volume of the solution increases. Recall that the concentration of a substance is defined as the number of molecules of that substance per unit volume of solution. In Figure 2-1, this means that the concentration of water in the sugar–water solution is lower than it was in the pure water before the sugar was dissolved. This is because the total volume increased after the sugar

Figure 2-1
When sugar molecules (filled circles) are dissolved in a liter of water, the resulting solution occupies a volume greater than a liter. This is because the sugar molecules have taken up some space formerly occupied by water molecules (open circles). Therefore, the concentration of water (number of molecules of water per unit volume) is lower in the sugar–water solution.

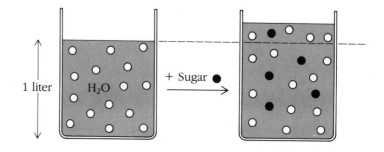

was added, but the total number of water molecules present is the same before and after dissolving the sugar in the water.

In order to compare the concentrations of water in solutions containing different concentrations of dissolved substances, we will use the concept of **osmolarity**. A solution containing 1 mole of dissolved particles per liter of solution (a 1 molar, or 1 M, solution) is said to have an osmolarity of 1 osmolar (1 Osm), and a 1 millimolar (1 mM) solution has an osmolarity of 1 milliosmolar (1 mOsm). *The higher the osmolarity of a solution, the lower the concentration of water.* For practical purposes in biological solutions, it doesn't matter what the dissolved particle is; that is, the concentration of water is effectively the same in a solution of 0.1 Osm glucose, 0.1 Osm sucrose, or 0.1 Osm urea. To be strictly correct in discussing the concentration of water in various solutions, we would have to speak of the **molality**, rather than the molarity, of the solutions. Whereas molarity is defined as moles of solute per liter of solution, molality is defined as moles of solute per kilogram of solvent. This means that molality takes into account the fact that higher molecular weight solutes displace more water per mole of solute than do lower molecular weight solutes. That is, a liter of solution containing 1 mole of a large molecule, like a protein, would contain less water than a liter of solution containing 1 mole of a small molecule, like urea. Thus, the molality of the protein solution would be higher than the molality of the urea solution, even though both solutions have the same molarity (1 M). For our purposes in this book, however, it will be adequate to treat molarity and osmolarity as equivalent to molality and osmolality.

It is important in determining the osmolarity of a solution to take into account how many dissolved particles result from each molecule of the dissolved substance. Glucose, sucrose, and urea molecules don't dissociate when they dissolve, and thus a 0.1 M glucose solution is a 0.1 Osm solution. A solution of sodium chloride,

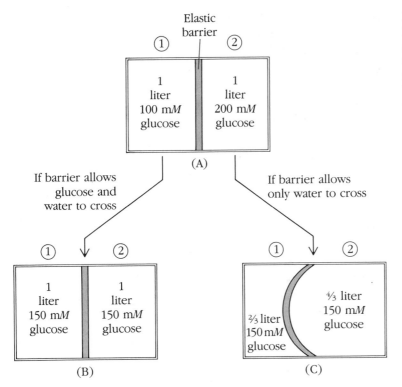

Figure 2-2
Effect of properties of the barrier separating two different glucose solutions on final volumes of the solutions. (A) If the barrier allows both glucose and water to cross, the volumes of the two solutions do not change when equilibrium is reached (B). If the barrier allows only water to cross, osmolarities of the two solutions are again the same at equilibrium, but the final volumes are different (C).

however, contains two dissolved particles—a sodium and a chloride ion—from each molecule of salt that goes into solution. Thus, a 0.1 M NaCl solution is a 0.2 Osm solution. To be strictly correct, we would have to take into account interactions among the ions in a solution, so that the effective osmolarity might be less than we would expect from assuming that all dissolved particles behave independently. But for dilute solutions like those we usually encounter in cell biology, such interactions are weak and can be safely ignored. Thus, for practical purposes we will assume that all dissolved particles act independently in determining the total osmolarity of a solution. This means that solutions containing 300 mM glucose, 150 mM NaCl, 100 mM NaCl + 100 mM glucose, or 75 mM NaCl + 75 mM KCl would all have the same total osmolarity—300 mOsm.

When solutions of different osmolarity are placed in contact through a barrier that allows water to move across, water will diffuse across the barrier down its concentration gradient (that is, from the lower osmolar solution to the higher). This movement of water down its concentration gradient is called osmosis. Consider the example shown in Figure 2-2A, which shows a container divided into two equal compartments that are filled with glucose solutions. Imagine that the barrier dividing the container is made of an elastic material, so that it can stretch freely. If the barrier allows both water and glucose to cross, then water will move from side 1 to side 2, down its concentration gradient, and glucose will move from side 2 to side 1. The movement of water and glucose will continue until their concentrations on the two sides of the barrier are equal. Thus, side 1 gains glucose and loses water, and side 2 loses glucose and gains water until the glucose concentration on both sides is 150 mM. There will be no net change in the volume of solution on either side of the barrier, as shown in Figure 2-2B.

If the barrier in Figure 2-2A allows water but not glucose to cross, however, the outcome will be quite different from that shown in Figure 2-2B. Once again, water will move down its concentration gradient from side 1 to side 2. In this case, though, the loss of water will not be compensated by a gain of glucose. As water continues to leave side 1 and accumulates on side 2, the volume of side 2 will increase and the volume of side 1 will decrease. The accumulating water will exert a pressure on the elastic barrier, causing it to expand to the left to accommodate the volume changes (as shown in Figure 2-2C). The resulting volume changes will increase the osmolarity of side 1 and decrease the osmolarity of side 2, and this will continue until the osmolarities of the two sides are equal—150 mOsm. In order to prevent the changes in volume, we would have to exert a pressure against the elastic barrier from side 1 to keep it from stretching. This pressure would be equal to the pressure moving water down its concentration gradient and would provide a measure of the **osmotic pressure** gradient across the barrier.

Osmotic Balance and Cell Volume

Return now to the hypothetical primitive cell, early after the development of a cell membrane. In order for the cell membrane to do its job, it must be impermeable to the organic molecules inside the cell. But if the compositions of the extracellular and intracellular fluids are the same, with the exception of the internal organic molecules, the cell faces an imbalance of water on the two sides of the membrane. This is shown schematically in Figure 2-3. Here, the solutes that are in common in ICF and ECF are grouped together and symbolized by S. The extra solute inside the cell—the organic molecules (symbolized by P, for protein)—cause the con-

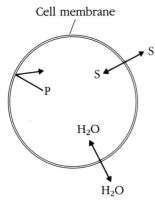

Cell membrane

Figure 2-3
A simple model cell containing organic molecules, P. The extracellular fluid is a solution of solute, S, in water. Both water and S can cross the cell membrane, but P cannot.

centration of water inside the cell to be less than it is outside. Put another way, the total osmolarity inside the cell is greater than it is outside the cell. There are two solutes inside, S and P, and only one outside. Water will therefore enter the cell and will continue to enter until the osmolarity on the two sides of the membrane is the same. Because the volume of the sea is essentially infinite relative to the volume of a cell and can thus be treated as constant, this end point could be reached only when the internal concentration of organic solutes is zero. This would require the volume of the cell to be infinite. Real cell membranes are not infinitely elastic, and thus water will enter the cell, causing it to swell, until the membrane ruptures and the cell bursts.

It will be convenient to summarize this situation in equation form. *If a substance is at diffusion equilibrium across a cell membrane, there is no net movement of that substance across the membrane.* For any solute, S, that can cross the cell membrane, diffusion equilibrium will be reached when

$$[S]_i = [S]_o \tag{2-1}$$

The square brackets indicate the concentration of a substance, and the subscripts i and o refer to the inside and outside of the cell. Thus, in order for water to be at equilibrium, we would expect that

$$[S]_i + [P]_i = [S]_o \tag{2-2}$$

which is the same as saying that at equilibrium, the total osmolarity inside the cell must be the same as the total osmolarity outside the cell. For the cell of Figure 2-3, diffusion equilibrium will be reached only when the concentrations of all substances that can cross the membrane (in this case, S and water) are the same inside and outside the cell. This would require that equations (2-1) and (2-2) be true simultaneously, which can occur only if $[P]_i$ is zero.

Answers to the Problem of Osmotic Balance

What solutions exist to this apparently fatal problem? There are three basic strategies that have developed in different types of cells. First, the problem could be eliminated by making the cell membrane impermeable to water. This turns out to be quite difficult to do and is not a commonly found solution to the problem of os-

Figure 2-4
Effects of various extracellular fluids on the volume of a simple model. (A) The ECF contains an impermeant solute (sucrose), and the osmolarity is the same as that inside the cell. (B) The ECF contains an impermeant solute, and the osmolarity is lower than that inside the cell. (C) The ECF contains a permeant solute (urea) and external and internal osmolarities are equal. (D) The ECF contains a mixture of permeant and impermeant solutes.

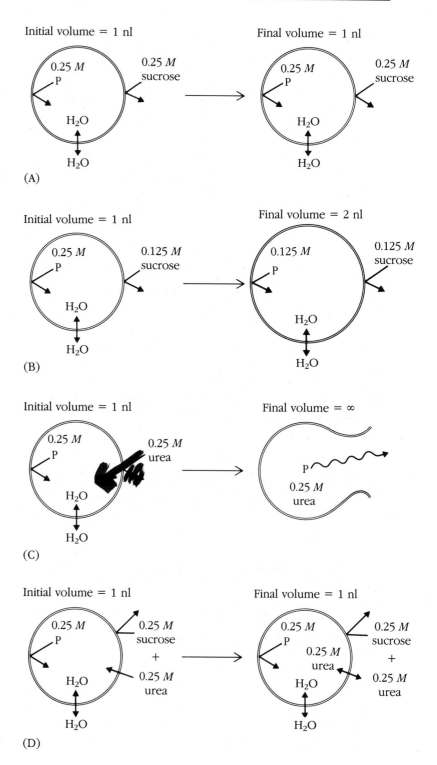

motic balance. However, certain kinds of epithelial cells have achieved very low permeability to water. A second strategy is commonly found and was likely the first solution to the problem. Here, the basic idea is to use brute force: build an inelastic wall around the cell membrane to physically prevent the cell from swelling. This is the solution that is used by bacteria and plants. The third strategy is that found in animal cells: achieve osmotic balance by making the cell membrane impermeable to selected extracellular solutes. This solution to the problem of osmotic balance works by balancing the concentration of nonpermeating molecules inside the cell with the same concentration of nonpermeating solutes outside the cell.

To see how the third strategy works, it will be useful to work through some examples using a simplified model animal cell whose membrane is permeable to water. Suppose that the model cell contains only one solute: nonpermeating protein molecules, P, dissolved in water at a concentration of 0.25 M. We will then perform a series of experiments on this model cell by placing it in various extracellular fluids and deducing what would happen to its volume in each case. Assume that the initial volume of the cell is one billionth of a liter (1 nanoliter, or 1 nl) and that the volume of the extracellular fluid in each case is infinite. This latter assumption means that the concentration of extracellular solutes does not change during the experiments.

The first experiment will be to place the cell in a 0.25 M solution of sucrose, which does not cross cell membranes. This is shown in Figure 2-4A. In this situation, only water can cross the cell membrane. For water to be at equilibrium, the internal osmolarity must equal the external osmolarity, or:

$$[P]_i = [sucrose]_o \tag{2-3}$$

Because the internal and external osmolarities are both 0.25 Osm, this condition is met; there will be no net diffusion of water, and cell volume will not change.

In the second example, shown in Figure 2-4B, imagine that the cell is placed in 0.125 M sucrose rather than 0.25 M sucrose. Again, only water can cross the membrane, and equation (2-3) must be satisfied for equilibrium to be reached. In 0.125 M sucrose, however, the internal osmolarity is greater than the external, and water will enter the cell until internal osmolarity falls to 0.125 M. This will happen when the cell volume is twice normal, that is, 2 nl. What would the equilibrium cell volume be if we placed the cell in 0.5 M sucrose rather than 0.125 M?

The point of the previous two examples is that *water will be at equilibrium if the concentration of impermeant extracellular solute is the same as the concentration of impermeant internal solute.* To see that the external solute must not be able to cross the cell membrane, consider the example shown in Figure 2-4C. In this case, the model cell is placed in 0.25 M urea, rather than sucrose. Unlike sucrose, urea can cross the cell membrane, and thus we must take into account both urea and water in determining diffusion equilibrium. In equation form, equilibrium will be reached when these two relations hold:

$$[urea]_i = [urea]_o \qquad (2\text{-}4)$$

$$[urea]_i + [P]_i = [urea]_o \qquad (2\text{-}5)$$

Because the external volume is infinite, $[urea]_o$ will be 0.25 M at equilibrium, and according to equation (2-4) $[urea]_i$ will also be 0.25 M. Equations (2-4) and (2-5) require, then, that $[P]_i$ be zero at equilibrium. Thus, the equilibrium volume is infinite, and the cell will swell until it bursts. Qualitatively, when the cell is first placed in 0.25 M urea, there will be no net movement of water across the membrane because internal and external osmolarities are both 0.25 Osm. But as urea enters the cell down its concentration gradient, internal osmolarity rises as urea accumulates. Water will then begin to enter the cell down its concentration gradient. The cell begins to swell and continues to do so until it bursts. Thus, an extracellular solute that can cross the cell membrane cannot help a cell achieve osmotic balance.

An interesting example is shown in Figure 2-4D. In this experiment, the model cell is placed in a mixture of 0.25 M urea and 0.25 M sucrose. The equilibrium for urea will once again be given by equation (2-4), and water will be at equilibrium when

$$[urea]_i + [P]_i = [urea]_o + [sucrose]_o \qquad (2\text{-}6)$$

Both equation (2-4) and equation (2-6) will be satisfied when $[P]_i$ = 0.25 M, which is the initial condition. Therefore, in this example, the cell volume at diffusion equilibrium will be the normal volume, 1 nl. The point is that even if some extracellular solutes can cross the cell membrane, the presence of a nonpermeating external solute at the same concentration as the nonpermeating internal solute allows the cell to achieve diffusion equilibrium for water and thus to maintain its volume. This is the strategy taken by animal cells to avoid bursting. As shown in Table 1-1, the impermeant extracellular solute is sodium.

In all the examples of osmotic equilibrium we just worked through, the answer was arrived at using just one rule: *For each permeating substance (including water), the inside concentration must equal the outside concentration at equilibrium.*

Tonicity

In the examples in Figure 2-4, 0.25 *M* sucrose and 0.25 *M* urea had the same osmolarity: 0.25 Osm. But the two solutions had dramatically different effects on cell volume. In 0.25 *M* sucrose, cell volume didn't change, while in 0.25 *M* urea the cell exploded. To take into account the differing biological effects of solutions of the same osmolarity, we will use the concept of **tonicity**. An **isotonic** solution has no final effect on cell volume; a solution that causes cells to swell at equilibrium is called a **hypotonic** solution; and a solution that causes cells to shrink at equilibrium is called a **hypertonic** solution. Thus, the 0.25 *M* sucrose solution was isotonic, and the 0.25 *M* urea solution was hypotonic. Note that an isotonic solution must have the same osmolarity as the fluid inside the cell, but that having the same osmolarity as the intracellular fluid does not guarantee that an external fluid is isotonic.

Time-Course of Volume Changes

So far in the discussion of maintenance of cell volume, we have considered only the final, equilibrium effect of a solution on cell volume and have ignored any transient effects that may occur. To see such transient effects, consider what happens to the model cell immediately after it is placed in the solution in Figure 2-4D, 0.25 *M* urea + 0.25 *M* sucrose. This is summarized in Figure 2-5. At the start, the osmolarity outside (0.5 Osm) is greater than the osmolarity inside (0.25 Osm), and water will initially leave the cell as it diffuses down its concentration gradient. Urea, however, begins to diffuse into the cell down its concentration gradient. Thus, the

Figure 2-5

Time-course of cell volume when the model cell is placed in solution (D) of Figure 2-4.

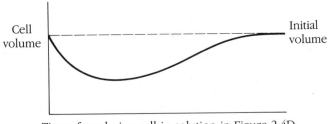

Time after placing cell in solution in Figure 2-4D

internal osmolarity begins to rise as a result of the increasing $[urea]_i$ and the loss of intracellular water. The leakage of water out of the cell slows down and finally ceases altogether when $[P]_i + [urea]_i = 0.5\ M$; that is, at the point when internal and external osmolarities are equal. At this point, however, $[P]_i$ is higher than its initial value $(0.25\ M)$ because of the reduction in cell volume, and $[urea]_i$ is thus less than $0.25\ M$. Urea therefore continues to enter the cell to reach its own diffusion equilibrium, and the internal osmolarity rises above 0.5 Osm, so that water enters the cell and volume begins to increase. This situation continues until the final equilibrium state governed by equations (2-4) and (2-6) is reached. What would you expect the time-course of cell volume to be if the model cell were placed in an infinite volume of a solution of $0.5\ M$ urea?

Summary

If animal cells are to survive, it is essential that they regulate the movement of water across the plasma membrane. Given that proteins and other organic constituents of the ICF cannot be allowed to cross the membrane, diffusion of water becomes a problem. Animal cells have solved this problem by excluding a compensating extracellular solute, sodium ions. We'll discuss in more detail later exactly how they go about excluding Na^+.

Diffusion equilibrium is reached when internal and external concentrations are equal for all substances that can cross the membrane. For uncharged substances, such as those we have considered in our examples so far, we do not have to consider the influence of electrical force on the equilibrium state. However, the solutes of the ICF and ECF of real cells bear a net electrical charge. In the next chapter, we will consider what role electric fields play in the movements of these charged substances across the membranes of animal cells.

3

Ionic Equilibrium and Membrane Potential

The central topics in Chapter 2 were the factors that influence the distribution of water across the plasma membrane and the strategies by which cells can attain osmotic equilibrium. For clarity, all the examples so far have used only uncharged particles; however, a glance at Table 1-1 in Chapter 1 shows that all the solutes of both ICF and ECF are electrically charged. For charged particles, movement across the membrane will be determined not only by diffusional force, but also by the electrical potential across the membrane. This chapter will consider how cells can achieve equilibrium in the situation where both diffusional and electrical forces must be taken into account.

To illustrate the important principles that apply to ionic equilibrium, it will be useful to work through a series of examples that are increasingly complex and increasingly similar to the situation in real animal cells. At the end of the series of examples, we will see how a model cell, with internal and external compositions like those given in Table 1-1, could be in electrical and chemical equilibrium. However, we will also see that this equilibrium model of the electrochemical state of cells does not apply to real animal cells, and we will see that real cells must expend energy to maintain the distribution of ions across their plasma membranes.

Diffusion Potential

The names **cation** for positively charged particles and **anion** for negatively charged ions arise from the observation that dissolved

Figure 3-1

Schematic diagram of apparatus for measuring diffusion potential. The voltmeter is wired to measure the electrical voltage difference across the barrier separating the two salt solutions.

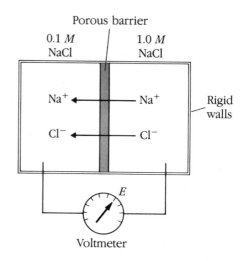

Porous barrier

0.1 *M* NaCl   1.0 *M* NaCl

Na⁺ ← Na⁺

Rigid walls

Cl⁻ ← Cl⁻

E

Voltmeter

positively charged particles accumulate around a wire connected to the negative pole, or cathode, of a battery and that dissolved negatively charged particles are attracted to a wire connected to the positive pole, or anode. The battery sets up a gradient of electrical potential (a **voltage** gradient) in the solution, and the movement of the ions in the solution is influenced by that voltage gradient. Thus, the distribution of ions in a solution depends on the presence of an electric field in that solution. The other side of the coin is that a differential distribution of ions in a solution *gives rise to* a voltage gradient in the solution. As an example of how an electrical potential can arise from spatial differences in the distribution of ions, we will consider the origin of **diffusion potentials**.

Diffusion potentials arise in the situation where two or more ions are moving down a concentration gradient. Examine the situation illustrated in Figure 3-1, which shows a rigid container divided into two compartments by a porous barrier. In the left compartment we place a 0.1 *M* NaCl solution and in the right compartment a 1.0 *M* NaCl solution. The porous barrier allows Na⁺, Cl⁻, and water to cross, but because of the rigid walls the compartment volume is not free to change and water cannot move. Thus, osmotic factors can be neglected for the moment. However, both Na⁺ and Cl⁻ will move down their concentration gradients from right to left until their concentrations are equal in both compartments. In aqueous solution, Na⁺ and Cl⁻ do not move at the same rate; Cl⁻ is more mobile and moves from right to left more quickly than Na⁺. This is because ions dissolved in water carry with them a loosely associated "cloud" of water molecules, and Na⁺ must drag along a larger cloud than Cl⁻, causing it to move more slowly.

In Figure 3-1, then, the concentration of Cl^- on the left side will rise faster than the concentration of Na^+. In other words, there will be more negative than positive charges in the left compartment, and a voltmeter connected between the two sides would record a voltage difference, E, across the barrier, with the left compartment being negative with respect to the right compartment. This voltage difference is the diffusion potential. Notice that the electrical potential across the barrier tends to retard movement of Cl^- and speed up movement of Na^+ because the excess negative charges on the left repel Cl^- and attract Na^+. The diffusion potential will continue to build up until the electrical effect on the ions exactly counteracts the greater mobility of Cl^-, and the two ions cross the barrier at the same rate.

Another name for voltage is electromotive force. This name emphasizes the fact that voltage is the driving force for the movement of electrical charges through space; without a voltage gradient there is no net movement of charged particles. Thus, voltage can be thought of as a pressure driving charges in a particular direction, just as the pressure in the water pipe drives water out through your tap when you open the valve. Unlike the pressure in a hydraulic system, however, a voltage gradient can move charges in two opposing directions, depending on the polarity of the charge. Thus, the negative pole of a battery simultaneously attracts positively charged particles and repels negatively charged particles.

Equilibrium Potential

The Nernst Equation

The diffusion potential example of Figure 3-1 does not describe an equilibrium condition, but rather a transient situation that occurs only as long as there is a net diffusion of ions across the barrier. Equilibrium would be achieved in Figure 3-1 only when $[Na^+]$ and $[Cl^-]$ are the same in compartments 1 and 2. At that point, there would be no concentrational force to support net diffusion of either Na or Cl across the membrane and there would be no electrical potential across the barrier. Under what conditions might there be a steady electrical potential at equilibrium? To see this, consider a small modification to the previous example, shown in Figure 3-2. In the new example, everything is as before, except that the barrier between the two compartments of the box is selectively permeable to Cl^-: Na^+ cannot cross. Once again, we assume that

Figure 3-2

Schematic diagram of apparatus for measuring the equilibrium, or Nernst, potential for a permeant ion. At equilibrium, there will be a steady electrical potential (the equilibrium potential) across the selectively permeable barrier separating the two salt solutions.

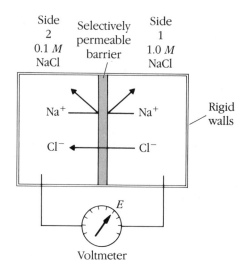

the box has rigid walls so that we can neglect movement of water for the present.

The analysis of the situation in Figure 3-2 is similar to that of the diffusion potential example, except that now the "mobility" of Na^+ is reduced effectively to zero by the permeability characteristics of the barrier. Chloride ions will move down their concentration gradient from compartment 1 to compartment 2, but now no positive charges accompany them and negative charges will quickly build up in compartment 2. Thus, the voltmeter will record an electrical potential across the barrier, with side 2 being negative with respect to side 1. Because only Cl^- can cross the barrier, equilibrium will be reached when there is no further net movement of chloride across the barrier. This happens when the electrical force driving Cl^- out of compartment 2 exactly balances the concentrational force driving Cl^- out of compartment 1. Thus, at equilibrium a chloride ion moves from side 1 to side 2 down its concentration gradient for every chloride ion that moves from side 2 to side 1 down its electrical gradient. There will be no further change in $[Cl^-]$ in the two compartments, and no further change in the electrical potential, once this equilibrium has been reached.

Equilibrium for an ion is determined not only by concentrational forces but also by electrical forces. Movement of an ion across a cell membrane is determined both by the concentration gradient for that ion across the membrane and by the electrical potential difference across the membrane. We will use these ideas extensively in this book, so the remainder of this chapter will be spent examining how these principles apply in simple model situations and in real cells.

What would be the measured value of the voltage across the barrier at equilibrium in Figure 3-2? This is a quantitative question, and the answer is provided by equation (3-1), which is called the **Nernst equation** after the physical chemist who derived it. The Nernst equation for Figure 3-2 can be written

$$E_{Cl} = (RT/ZF) \ln([Cl^-]_1/[Cl^-]_2) \qquad (3\text{-}1)$$

Here, E_{Cl} is the voltage difference between sides 1 and 2 at equilibrium, R is the gas constant, T is the absolute temperature, Z is the valence of the ion in question (-1 for chloride), F is Faraday's constant, ln is the symbol for the natural, or base e, logarithm, and $[Cl^-]_1$ and $[Cl^-]_2$ are the chloride concentrations in compartments 1 and 2.

The value of electrical potential given by equation (3-1) is called the **equilibrium potential**, or **Nernst potential**, for the ion in question. For example, in Figure 3-2 the permeant ion is chloride and the electrical potential, E_{Cl}, across the barrier is called the chloride equilibrium potential. If the barrier in Figure 3-2 allowed Na^+ to cross rather than Cl^-, equation (3-1) would again apply, except that $[Na^+]_1$ and $[Na^+]_2$ would be used instead of $[Cl^-]$. In the latter case, the resulting potential, E_{Na}, would be the sodium equilibrium potential.

The Nernst equation applies only to one ion at a time and only to ions that can cross the barrier.

A derivation of equation (3-1) is given in Appendix A. The Nernst equation comes from the realization that at equilibrium the total change in energy encountered by an ion in crossing the barrier must be zero. If the change in energy were not zero, there would be a net force driving the ion in one direction or the other, and the ion would not be at equilibrium. There are two important sources of energy change involved in crossing the barrier shown in Figure 3-2: the electric field and the concentration gradient. Nernst arrived at his equation by setting the sum of the concentrational and electrical energy changes across the barrier to zero.

In biology, we usually work with a simplified form of equation (3-1):

$$E_{Cl} = (58 \text{ mV}/Z) \log([Cl^-]_1/[Cl^-]_2) \qquad (3\text{-}2)$$

The simplification arises from converting from natural to base 10 logarithms, evaluating (RT/F) at standard room temperature (20°C), and expressing the result in millivolts (mV). That is where the constant 58 mV comes from in equation (3-2). From the simplified Nernst equation, it can be seen that E_{Cl} in Figure 3-2 would be -58 mV. That is, in crossing the barrier from side 1 to side 2, we would encounter a potential change of 58 mV, with side 2 being negative with respect to side 1. This is as expected from the fact that chloride ions, and therefore negative charges, are accumulating on side 2. If the barrier were selectively permeable to Na^+ rather than Cl, the voltage across the barrier would be given by E_{Na}, which would be $+58$ mV given the values in Figure 3-2. What would be the equilibrium potential for chloride in Figure 3-2 if the concentration of NaCl was $1.0\,M$ on both sides of the barrier? (Hint: in that case the concentration gradient would be zero.)

The Principle of Electrical Neutrality

In arriving at -58 mV for the chloride equilibrium potential in Figure 3-2, we used $1.0\,M$ and $0.1\,M$ for $[Cl^-]_1$ and $[Cl^-]_2$. These are the initial concentrations in the two compartments, even though in our qualitative analysis we said that Cl^- moved from compartment 1 to 2, producing an excess of negative charge in compartment 2 and giving rise to the electrical potential. This would seem to suggest that $[Cl^-]$ changes from its initial value, invalidating our sample calculation. It is legitimate to use initial concentrations, however, because the increment in the electrical gradient caused by the movement of a single charged particle from compartment 1 to 2 is very much larger than the decrement in concentration gradient resulting from movement of that same particle. Thus, *only a very small number of charges need accumulate in order to counter even a large concentration gradient.*

In Figure 3-2, for example, it is possible to calculate that if the volume of each compartment were 1 ml and if the barrier between compartments were 1 cm² of the same material as found in cell membranes, it would require less than 1 billionth of the chloride ions of side 1 to move to side 2 in order to reach the equilibrium potential of -58 mV. (The basis of this calculation is explained below.) Clearly, such a small change in concentration would produce an insignificant difference in the result calculated according to equation (3-2), and we can safely ignore the movement of chloride necessary to achieve equilibrium.

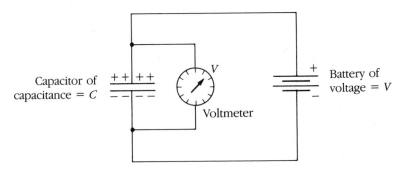

This leads to an important principle that will be useful in the examples following in this chapter. This principle, called the **principle of electrical neutrality**, states that *under biological conditions the bulk concentration of cations within any compartment must be equal to the bulk concentration of anions in that compartment*. This is an acceptable approximation because the number of charges necessary to reach transmembrane potentials of the magnitude encountered in biology is insignificant compared with the total numbers of cations and anions in the intracellular and extracellular fluids.

The Cell Membrane as an Electrical Capacitor

This section explains how we were able to calculate the number of charges necessary to produce the equilibrium membrane potential of -58 mV in the preceding section. The calculation was made by treating the barrier between the two compartments as an electrical capacitor, which is a charge-storing device consisting of two conducting plates separated by an insulating barrier. In Figure 3-2, the two conducting plates are the salt solutions in the two compartments, and the barrier is the insulator. In a real cell, the ICF and ECF are the conductors, and the lipid bilayer of the plasma membrane is the insulating barrier. When a capacitor is hooked up to a battery as shown in Figure 3-3, the voltage of the battery causes electrons to be removed from one conducting plate and to accumulate on the other plate. This will continue until the resulting voltage gradient across the capacitor is equal to the voltage of the battery. Basic physics tells us that the amount of charge, q, stored on the capacitor at that time will be given by $q = CV$, where V is the voltage of the battery and C is the **capacitance** of the capacitor. A capacitor's capacitance is directly proportional to the area of the plates (bigger plates can store more charge) and inversely proportional to the distance separating the two plates. Capacitance also depends on the characteristics of the insulating material between the plates; in the case of cells, that insulating material is the lipid plasma membrane. The unit of capacitance is the farad (F); a 1 F capacitor can store 1 coulomb of charge when hooked up to a 1 V battery. Biological membranes, like the plasma membrane, have a capacitance of 10^{-6} F (that is, 1 microfarad, or μF) per cm^2 of membrane area.

If the barrier in Figure 3-2 were 1 cm^2 of cell membrane, it would therefore have a capacitance of 10^{-6} F. From $q = CV$, it follows that an equilibrium potential of -58 mV would store 5.8×10^{-8} coulomb of charge on the barrier. Note that the charge on the membrane barrier in Figure 3-2 is carried by ions, not by electrons as in Figure 3-3. Thus, to know the total number of excess anions on side 2 of the barrier at equilibrium, we must convert from coulombs of charge to moles of ion. This can be done by dividing the number of coulombs on the barrier by Faraday's constant (approximately 10^5 coulombs per mole of monovalent ion), yielding 5.8×10^{-13} mole or about 3.5×10^{11} chloride ions moving from side 1 to side 2 in Figure 3-2. If the volume of each compartment were 1 ml, then side 2 would contain about 6×10^{20} chloride and sodium ions. This leads to the conclusion stated in the previous section that less than one billionth of the chloride ions in side 1 cross to side 2 to produce the equilibrium voltage across the barrier.

Incorporating Osmotic Balance

The example shown in Figure 3-2 illustrates how ionic equilibrium can be reached and how the Nernst equation can be used to calculate the value of the membrane potential at equilibrium. However, the simple situation in the example is not very similar to the situation in real animal cells. For one thing, animal cells are not enclosed in a box with rigid walls, and thus osmotic balance must be taken into account. An example of how equilibrium can be reached when water balance must be considered is shown in Figure 3-4A. In this example the rigid walls are removed, so that osmotic balance must be achieved in order to reach equilibrium. In addition, an impermeant intracellular solute, P, has been added. For now, P has no charge; the effect of adding a charge on the intracellular organic solute will be considered later in this chapter.

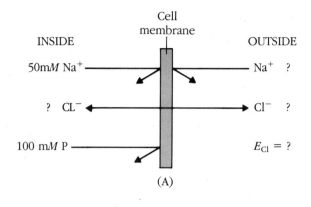

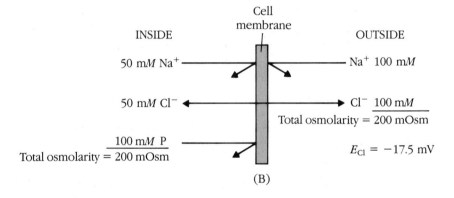

Figure 3-4
A model cell in which both osmotic and electrical factors must be considered at equilibrium.

In Figure 3-4A, it is assumed that the model cell contains 50 mM Na$^+$ and 100 mM P. What must the concentrations of the other intracellular and extracellular solutes be in order for the model cell to be at equilibrium? The principal of electrical neutrality tells us that for practical purposes, the concentrations of cations and anions within any compartment are equal. Thus, because P is assumed to have no charge, $[Cl^-]_i = [Na^+]_i = 50$ mM. For osmotic balance, the external osmolarity must equal the internal osmolarity, which is 200 mOsm. The principal of electrical neutrality again requires that $[Na^+]_o = [Cl^-]_o$. This requirement, together with the requirement for osmotic balance, can be satisfied if $[Na^+]_o = [Cl^-]_o = 100$ mM. The model cell of Figure 3-4A can therefore be at equilibrium if the concentrations of intracellular and extracellular solutes are as shown in Figure 3-4B. At this equilibrium, the electrical potential across the membrane of the model cell (the membrane potential, E_m) would be given by the Nernst equation for chloride:

$$E_m = E_{Cl} = -58 \text{ mV} \log([Cl^-]_o/[Cl^-]_i) = -17.5 \text{ mV.}$$

Donnan Equilibrium

The example of Figure 3-4B shows how we could construct a model cell that is simultaneously at osmotic and ionic equilibrium. However, the situation in Figure 3-4B is not very much like that in real animal cells. A major difference is that the principal internal cation in real cells is K^+, not Na^+. Also, there is some potassium in the ECF, and the cell membrane is permeable to K^+ as well as Cl^-. In this situation, there are two ions that can cross the membrane: K^+ and Cl^-. If equilibrium is to be reached, the electrical potential across the cell membrane must exactly balance the concentration gradients for both K^+ and Cl^-. Because the membrane potential can have only one value, this equilibrium condition will be satisfied only when the equilibrium potentials for Cl^- and K^+ are equal. In equation form, this condition can be written

$$E_K = 58 \text{ mV } \log([K^+]_o/[K^+]_i) =$$

$$E_{Cl} = -58 \text{ mV } \log([Cl^-]_o/[Cl^-]_i)$$

Here, the minus sign on the far right arises from the fact that the valence of chloride is -1. Cancelling 58 mV from the above relation leaves

$$\log ([K^+]_o/[K^+]_i) = -\log ([Cl^-]_o/[Cl^-]_i) \tag{3-3}$$

The minus sign on the right side can be moved inside the parentheses of the logarithm to yield $\log([Cl^-]_i/[Cl^-]_o)$. Thus, equilibrium will be reached when

$$([K^+]_o/[K^+]_i) = ([Cl^-]_i/[Cl^-]_o) \tag{3-4}$$

This equilibrium condition is called the **Donnan** or **Gibbs–Donnan equilibrium**, and it specifies the conditions that must be met if two ions that can cross a cell membrane are simultaneously to be at equilibrium across that membrane. Equation (3-4) is usually written in a slightly rearranged form as the product of concentrations:

$$[K^+]_o[Cl^-]_o = [K^+]_i[Cl^-]_i \tag{3-5}$$

In words, for a Donnan equilibrium to hold, *the product of the concentrations of the permeant ions outside the cell must be equal to the product of the concentrations of those two ions inside the cell.*

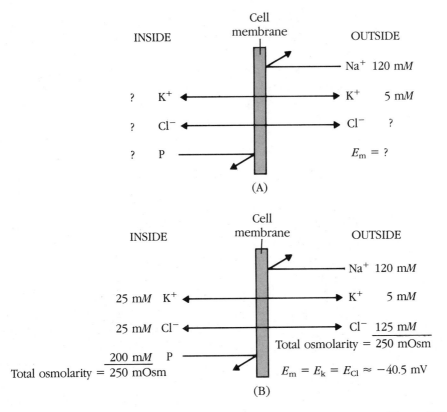

(A)

(B)

Figure 3-5

An example of a model cell at Donnan equilibrium. The cell membrane is permeable to both potassium and chloride.

To see how the Donnan equilibrium might apply in an animal cell, consider the example shown in Figure 3-5A. Here a model cell containing K^+, Cl^-, and P is placed in ECF containing Na^+, K^+, and Cl^-. As an exercise, we will calculate the values of all concentrations at equilibrium assuming that $[Na^+]_o$ is 120 mM and $[K^+]_o$ is 5 mM. From the principal of electrical neutrality, $[Cl^-]_o$ must be 125 mM. Also, because P is assumed for the present to be uncharged, the principle of electrical neutrality requires that $[K^+]_i$ must equal $[Cl^-]_i$. Because two ions—K and Cl—can cross the membrane, the defining relation for a Donnan equilibrium shown in equation (3-5) must be obeyed. Thus, if the model cell of Figure 3-5A is to be at equilibrium, $[K^+]_i[Cl^-]_i$ must equal $[K^+]_o[Cl^-]_o$, which is 5 × 125, or 625 mM^2. Because $[K^+]_i = [Cl^-]_i$, the Donnan condition reduces to $[K^+]_i^2 = 625$ mM^2; thus, $[K^+]_i$ and $[Cl^-]_i$ must be 25 mM at equilibrium. For osmotic balance, the internal osmolarity must equal the external osmolarity, which is 250 mOsm. This requires that $[P]_i$ must be 200 mM for the model cell to be at equilibrium. The results of this example are summarized in Figure 3-5B, which represents a model cell at equilibrium. What would be the membrane potential of this equilibrated model cell? The Nernst equation—equation (3-2)—tells us that the membrane potential for a cell at equilibrium with $[K^+]_o = 5$ mM and $[K^+]_i = 25$ mM is about -40.5 mV, inside negative. You should satisfy yourself that the Nernst equation for chloride yields the same value for membrane potential.

A Model Cell That Looks Like a Real Animal Cell

The model cell of Figure 3-5B still lacks many features of real animal cells. For instance, as Table 1-1 shows, the internal organic molecules are charged, and this charge must be considered in the balance between cations and anions required by the principle of electrical neutrality. Recall that the category of internal anions, A^-, actually represents a diverse group of molecules, including proteins, charged amino acids, and sulfate and phosphate ions. Some of these bear a single negative charge, others two, and some even three net negative charges. Taken as a group, however, the average charge per molecule is slightly greater than -1.2. Thus, the internal impermeant anions can be represented as $A^{-1.2}$.

In addition, the model cell of Figure 3-5B lacked Na^+ inside the cell, while real ICF does contain a small amount of sodium. Addition of these complicating factors leads to the model cell of Figure 3-6A, which now contains all the constituents shown in Table 1-1. If the cell of Figure 3-6A is to be at equilibrium, what concentrations of the various ions in ECF and ICF would be required, and what would be the transmembrane potential? To begin, we will take some values from Table 1-1 and determine what the remaining parameters must be for the cell to be at equilibrium. Assume that $[K^+]_o = 5$ mM, $[Na^+]_o = 120$ mM, $[Cl^-]_i = 5$ mM, and $[A^{-1.2}]_i = 108$ mM. (Actually, it is not necessary to assume the concentration of A; it could be calculated from the other parameters. For mathematical simplicity, however, we will assume that it is known from the start.) Because Cl^- is the sole external anion, the principle of electrical neutrality requires that $[Cl^-]_o$ be 125 mM. Both K^+ and Cl^- can cross the membrane, so that the conditions for a Donnan equilibrium—equation (3-5)—must be satisfied. This requires that $[K^+]_i = 125$ mM. The equilibrated value of $[Na^+]_i$ can then be obtained from the requirements for osmotic balance; $[Na^+]_i$ must be 12 mM if internal and external osmolarities are to be equal.

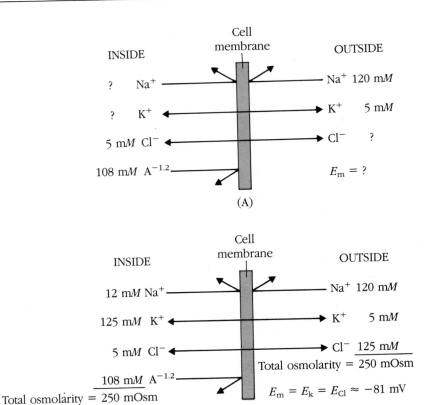

Cell membrane

INSIDE   OUTSIDE

? Na$^+$ —————— Na$^+$ 120 mM

? K$^+$ —————— K$^+$ 5 mM

5 mM Cl$^-$ —————— Cl$^-$?

108 mM A$^{-1.2}$ —————— E_m = ?

(A)

Cell membrane

INSIDE   OUTSIDE

12 mM Na$^+$ —————— Na$^+$ 120 mM

125 mM K$^+$ —————— K$^+$ 5 mM

5 mM Cl$^-$ —————— $\underline{\text{Cl}^- \ 125 \ mM}$

108 mM A$^{-1.2}$ —————— Total osmolarity = 250 mOsm

Total osmolarity = 250 mOsm $E_m = E_k = E_{Cl} \approx -81$ mV

(B)

Figure 3-6
An example of a realistic model cell that is at both electrical and osmotic equilibrium. The compositions of ECF and ICF for this equilibrated model cell are the same as for a typical mammalian cell (see Table 1-1).

From the Nernst equation for either Cl$^-$ or K$^+$, the membrane potential at equilibrium can be determined to be about −81 mV.

The equilibrium values for this model cell are shown in Figure 3-6B. Note that the concentrations of all intracellular and extracellular solutes are the same for the model cell and for real mammalian cells (Table 1-1). The values in Figure 3-6B were arrived at by assuming that the cell was in equilibrium, and this implies that the real cell, which has the same ECF and ICF, is also at equilibrium. Thus, the model cell, and by extension the real cell, will remain in the state summarized in Figure 3-6B without expending any metabolic energy at all. From this viewpoint, the animal cell is a beautiful example of efficiency, existing at perfect equilibrium, both ionic and osmotic, in harmony with its electrochemical environment. The problem, however, is that the model cell is not an accurate representation of the situation in real animal cells: *real cells are not at equilibrium and must expend metabolic energy to maintain the status quo.*

The Sodium Pump

For some time, the model in Figure 3-6B was thought to be an accurate description of real animal cells. The difficulty with this scheme arose when it became apparent that real cells are permeable to sodium, while the model cell is assumed to be impermeable to sodium. Permeability to sodium, however, would be catastrophic for the model cell. If sodium can cross the membrane, then all extracellular solutes can cross the membrane. Recall from Chapter 2, however, what happens to cells that are placed in ECF containing only permeant solutes (like the urea example in Figure 2-4C): the cell swells and bursts. The cornerstone of the strategy employed by animal cells to achieve osmotic balance is that the cell membrane must exclude an extracellular solute to balance the impermeant organic solutes inside the cell. Sodium ions played that role for the model cell of Figure 3-6B.

How can the permeability of the plasma membrane to sodium be reconciled with the requirement for osmotic balance? An answer to this question was suggested by the experiments that demonstrated the sodium permeability of the cell membrane in the first place. In these experiments, red blood cells were incubated in an external medium containing radioactive sodium ions. When the cells were removed from the radioactive medium and washed thoroughly, it was found that they remained radioactive, indicating that the cells had taken up some of the radioactive sodium. This showed that the plasma membrane was permeable to sodium. In addition, it was found that the radioactive cells slowly lost their radioactive sodium when incubated in normal ECF. This latter observation was surprising because both the concentration gradient and the electrical gradient for sodium are directed inward; neither would tend to move sodium out of the cell. Further, the rate of this loss of radioactive sodium from the cell interior was slowed dramatically by cooling the cells, indicating that a source of energy other than simple diffusion was being tapped to actively "pump" sodium out of the cell against its concentrational and electrical gradients. It turns out that this energy source is metabolic energy in the form of the high-energy phosphate compound adenosine triphosphate (ATP).

This active pumping of sodium out of the cell effectively prevents sodium from accumulating intracellularly as it leaks in down its concentration and electrical gradients. Thus, even though sodium can cross the membrane, it is actively extruded at a rate sufficiently high to counterbalance the inward leak. The net result is that *sodium behaves osmotically as though it cannot cross the mem-*

brane. Note however that this mechanism is fundamentally different from the situation in the model cell of Figure 3-6B. The model was in *equilibrium* and required no energy input to maintain itself. In contrast, real animal cells are in a finely balanced *steady state*, in which there is no net movement of ions across the cell membrane, but which requires the expenditure of metabolic energy.

Metabolic inhibitors, such as cyanide or dinitrophenol, prevent the pumping of sodium out of the cell and cause cells to gain sodium and swell. If ATP is added, the pump can operate once again and the accumulated sodium will be extruded. Similarly, other manipulations that reduce the rate of ATP production, like cooling, cause sodium accumulation and increased cell volume. Experiments of this type demonstrated the role of ATP in the active extrusion of sodium and the maintenance of cell volume. The mechanism of the sodium pump has been studied biochemically. The pump itself seems to be a particular kind of membrane-associated protein molecule that can bind both sodium ions and ATP at the intracellular face of the membrane. The protein then acts as an enzyme to cleave one of the high-energy phosphate bonds of the ATP molecule, using the released energy to drive the bound sodium out across the membrane by a process that is not yet completely understood.

The action of the sodium pump also requires potassium ions in the ECF. It is currently thought that binding of K^+ to a part of the protein on the outer surface of the cell membrane is required for the protein to return to the configuration in which it can again bind another ATP and sodium ions at the inner surface of the membrane. The potassium bound on the outside is released again on the inside of the cell, so that the protein molecule acts as a bidirectional pump carrying sodium out across the membrane and potassium in. Thus, the sodium pump is more correctly referred to as the sodium–potassium pump, and can be thought of as a shuttle carrying Na^+ out across the membrane, releasing it in the ECF, then carrying K^+ in across the membrane and releasing it in the ICF. Because the pump molecule splits ATP and binds both sodium and potassium ions, biochemists refer to this membrane-associated enzyme as a Na^+–K^+ ATPase.

Summary

The movement of charged substances across the plasma membrane is governed not only by the concentration gradient across the membrane but also by the electrical potential across the membrane. Equilibrium for an ion across the membrane is reached when the electrical gradient exactly balances the concentration gradient for that ion. The equation that expresses this equilibrium condition quantitatively is the Nernst equation, which gives the value of membrane potential that will exactly balance a given concentration gradient.

If more than one ion can cross the cell membrane, both can be at equilibrium only if the Nernst, or equilibrium, potentials for both ions are the same. This requirement leads to the defining properties of the Donnan, or Gibbs-Donnan, equilibrium, which applies simultaneously to two permeant ions. By working through a series of examples, we saw how it is possible to build a model cell that is at equilibrium and that has ICF, ECF, and membrane potential like that of real animal cells.

Real cells, however, were found to be permeable to sodium ions. This removed an important cornerstone of the equilibrated model cell, and forced a change in viewpoint about the relation between animal cells and their environment. Real cells must expend metabolic energy, in the form of ATP, in order to "pump" sodium out against its concentration and electrical gradients and thus to maintain osmotic balance. In the next chapter, we will consider what effect the sodium permeability of the plasma membrane might have on the electrical membrane potential. We will see how the membrane potential depends not only on the concentrations of ions on the two sides of the membrane, as in the Nernst equation, but also on the relative permeability of the membrane to those ions.

4

Ionic Steady State and Membrane Potential

In the last chapter, we saw that it was possible for two permeant ions to be at equilibrium across a cell membrane simultaneously. This Donnan equilibrium required that the membrane potential be equal to the Nernst potentials for the two ions. However, we also saw that real animal cells are permeable to sodium and thus that there are three major ions—potassium, chloride, and sodium—that can cross the plasma membrane. This chapter will be concerned with the effect of sodium permeability on membrane potential and with the quantitative relation between ionic permeabilities and concentrations on the one hand and electrical membrane potential on the other.

Equilibrium Potentials for Sodium, Potassium, and Chloride

If the permeability of the cell membrane to sodium is not zero, then the resting membrane potential of the cell must have a contribution from Na^+ as well as from K^+ and Cl^-. This is true even though the sodium pump eventually removes any sodium that leaks into the cell. There are two reasons for this. First, recall that electrical force per particle is much stronger than concentrational force per particle; therefore, even a tiny trickle of sodium that would cause a negligible change in internal concentration could produce large changes in membrane potential. Because the sodium pump responds only to changes in the bulk concentration of sodium inside the cell, it could not detect and respond to the tiny changes that would occur for even large changes in membrane potential. Second, even though sodium that leaks in is eventually pumped out, the efflux of sodium through the pump is coupled with an influx of potassium. Thus, there is a net transfer of positive charge into the cell associated with leakage of sodium.

Application of the Nernst equation to the concentrations of sodium, potassium, and chloride in the ICF and ECF of a typical mammalian cell (Table 1-1) shows that the membrane potential cannot possibly be simultaneously at the equilibrium potentials of all three ions. As we calculated in Chapter 3, $E_K = E_{Cl} =$ about -80 mV (actually a bit greater than -81 mV, given the values in Table 1-1). But with $[Na^+]_o = 120$ mM and $[Na^+]_i = 12$ mM, E_{Na} would be $+58$ mV. The membrane potential, E_m, cannot simultaneously be at -80 mV and $+58$ mV. The actual value of membrane potential will fall somewhere between these two extreme values. If the sodium permeability of the membrane were in fact zero, E_m would be determined solely by E_K and E_{Cl} and would be -80 mV. Conversely, if chloride and potassium permeability were zero, E_m would be determined only by sodium and would lie at E_{Na}, $+58$ mV. Because the permeabilities of all three ions are nonzero, there will be a struggle between Na^+ on the one hand, tending to make E_m equal $+58$ mV, and K^+ and Cl^- on the other, tending to make E_m equal -80 mV. Two factors determine where E_m will actually fall: (1) ion concentrations, which determine the equilibrium potentials for the ions; and (2) relative ion permeabilities, which determine the relative importance of a particular ion in governing where E_m lies. Before expressing these relations quantitatively, it will be useful to consider the mechanism of ionic permeability in more detail.

Ionic Channels in the Plasma Membrane

The permeability of a membrane to a particular ion is a measure of the ease with which that ion can cross the membrane. It is a property of the membrane itself. Recall that ions cannot cross membranes through the lipid portion of the membrane; they must cross through aqueous pores or channels in the membrane. Thus, the ionic permeability of a membrane is determined by the properties of the ionic pores or channels in that membrane. The total permeability of a membrane to a particular ion is governed by the total number of membrane channels that allow that ion to cross and by the ease with which the ion can go through a single channel. Ionic channels are protein molecules that are associated with the membrane, and thus an important function of membrane proteins is the regulation of ionic permeability of the cell membrane. In later chapters, we will discuss how specialized channels modulate ionic permeability in response to chemical or electrical signals and the role of such changes in permeability in the processing of signals in the nervous system.

Not all membrane channels allow all ions to cross with equal ease. Some channels allow only cations through, others only anions. Some channels are even more selective, allowing only K^+ through but not Na^+, or vice versa. Thus, it is possible for a membrane to have very different permeabilities to different ions, depending on the number of channels for each ion.

Membrane Potential and Ionic Permeability

As an example of how the actual value of membrane potential depends on the relative permeabilities of the competing ions, consider the situation illustrated in Figure 4-1. This model cell is much more permeable to K^+ than to Na^+. In other words, there are many channels that allow K^+ to cross the membrane but only a few that allow Na^+ to cross. Imagine that initially we connect the cell to an apparatus that artificially maintains the resting membrane potential at E_K, so that $E_m = E_K = -80$ mV. (This could be accomplished experimentally using a voltage clamp apparatus, as described in Chapter 6). What will happen to E_m when we switch off the apparatus and allow E_m to take on any value it wishes? In order to determine what will happen, it is necessary to keep in mind one important principle: *if the membrane potential is not equal to the equilibrium potential for an ion, that ion will move across the membrane in such a way as to force E_m toward the equilibrium potential for that ion.* For example, Figure 4-2 illustrates the movement of K^+ across a cell membrane in response to changes in E_m. In this example, a cell is connected to an apparatus that allows us to set the membrane potential to any value we choose. Initially, we set E_m to E_K. Recall from Chapter 3 that when $E_m = E_K$ there is a balance between the electrical force driving K^+ into the cell and

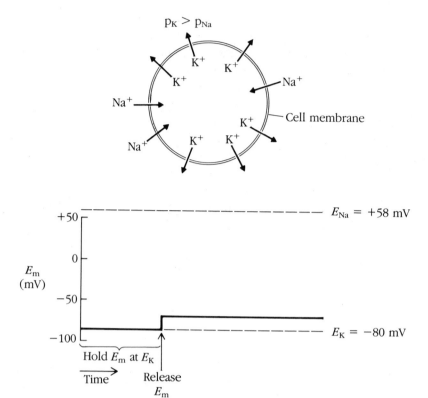

Figure 4-1

The resting membrane potential of a cell that is more permeable to potassium than to sodium. At the upward arrow, an apparatus that artificially holds the membrane potential at E_K is abruptly switched off, and E_m is allowed to seek its own resting level.

the concentrational force driving K^+ out of the cell. At time = a, however, we suddenly make the interior of the cell less negative, reducing the electrical potential across the cell membrane and therefore decreasing the electrical force driving K^+ into the cell. Such a reduction in the electrical potential across the membrane is called a depolarization of the membrane. The electrical force will then be weaker than the oppositely directed concentrational force, and there will be a net movement of K^+ out of the cell. Note that this movement is in the proper direction to make E_m move back toward E_K; that is, to make the interior of the cell more negative because of the efflux of positive charge. At time = b, we suddenly make E_m more negative than E_K; that is, we hyperpolarize the membrane. Now the electrical force will be stronger than the concentrational force and there will be a net movement of K^+ into the cell. Again, this is in the proper direction to make E_m move toward E_K, in this case by adding positive charge to the interior of the cell.

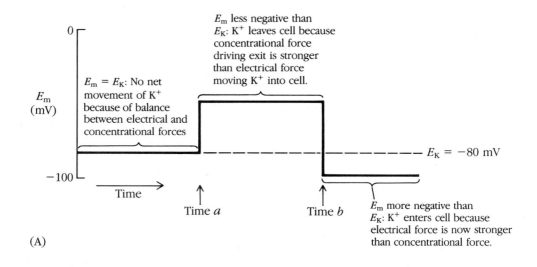

(A)

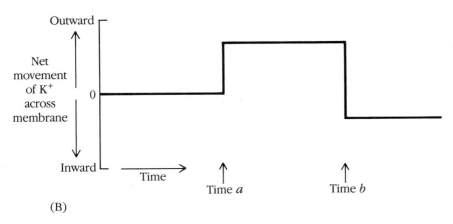

(B)

Return now to Figure 4-1. We would expect that Na^+, which has an equilibrium potential of $+58$ mV, will enter the cell. That is, Na^+ will bring positive charge into the cell, and when we switch off the apparatus forcing E_m to remain at E_K, this influx of sodium ions will cause the membrane potential to become more positive (that is, move toward E_{Na}). As E_m moves toward E_{Na}, however, it will no longer be equal to E_K, and K^+ will move out of the cell in response to the resulting imbalance between the potassium concentrational force and electrical force. Thus, there will be a struggle between K^+ efflux forcing E_m toward E_K and Na^+ influx forcing E_m toward E_{Na}. Because K^+ permeability is much higher than Na^+ permeability, potassium ions can move out readily to counteract the electrical effect of the trickle of sodium ions into the cell. Thus, in this

Figure 4-2

Effect of changes in membrane potential on the movement of potassium ions across the plasma membrane. When the membrane potential is artificially manipulated (A), potassium ions move across the membrane in a direction governed by the difference between E_m and E_K (B).

Figure 4-3
The resting membrane potential of a cell that is more permeable to sodium than to potassium. As in Figure 4-1, an apparatus holding E_m at E_K is abruptly turned off at the upward arrow.

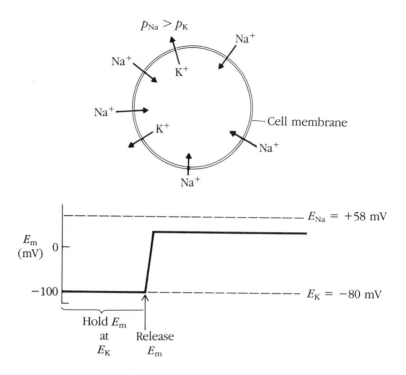

situation, the balance between the movement of Na^+ into the cell and the exit of K^+ from the cell would be struck relatively close to E_K.

Figure 4-3 shows a different situation. In this case, everything is as before except that the sodium permeability is much greater than the potassium permeability. That is, there are more channels that allow Na^+ across than allow K^+ across. Once again, we start with $E_m = E_K = -80$ mV and then allow E_m to seek its own value. Sodium, with $E_{Na} = +58$ mV, enters the cell down its electrical and concentration gradients. The resulting accumulation of positive charge again causes the cell to depolarize, as before. Now, however, potassium cannot move out as readily as sodium can move in, and the influx of sodium will not be balanced as readily by efflux of potassium. Thus, E_m will move farther from E_K and will reach a steady value closer to E_{Na} than to E_K.

The point of the previous two examples is that the value of membrane potential will be governed by the relative permeabilities of the permeant ions. *If a cell membrane is highly permeable to an ion, that ion can respond readily to deviations away from its equilibrium potential and E_m will tend to be near that equilibrium potential.*

The Goldman Equation

The examples discussed so far have been concerned with the qualitative relation between membrane potential and relative ionic permeabilities. The equation that gives the quantitative relation between E_m on the one hand and ion concentrations and permeabilities on the other is the **Goldman equation**, which is also called the **constant-field equation**. For a cell that is permeable to potassium, sodium and chloride, the Goldman equation can be written

$$E_m = \frac{RT}{F} \ln\left(\frac{p_k[K^+]_o + p_{Na}[Na^+]_o + p_{Cl}[Cl^-]_i}{p_k[K^+]_i + p_{Na}[Na^+]_i + p_{Cl}[Cl^-]_o}\right) \quad (4\text{-}1)$$

This equation is similar to the Nernst equation (see Chapter 3), except that it simultaneously takes into account the contributions of all permeant ions. Some information about the derivation of the Goldman equation can be found in Appendix B. Note that the concentration of each ion on the right side of the equation is scaled according to its permeability, p. Thus, if the cell is highly permeable to potassium, for example, the potassium term on the right will dominate and E_m will be near the Nernst potential for potassium. Note also that if p_{Na} and p_{Cl} were zero, the Goldman equation would reduce to the Nernst equation for potassium, and E_m would be exactly equal to E_K, as we would expect if the only permeant ion were potassium.

Because it is easier to measure relative ion permeabilities than it is to measure absolute permeabilities, the Goldman equation is often written in a slightly different form:

$$E_m = 58 \text{ mV } \log\left(\frac{[K^+]_o + b[Na^+]_o + c[Cl^-]_i}{[K^+]_i + b[Na^+]_i + c[Cl^-]_o}\right) \quad (4\text{-}2)$$

In this case, the permeabilities have been expressed relative to the permeability of the membrane to potassium. Thus, $b = p_{Na}/p_K$, and $c = p_{Cl}/p_K$. We have also evaluated RT/F at room temperature, converted from ln to log, and expressed the result in millivolts.

For most nerve cells, the Goldman equation can be simplified even further: the chloride term on the right can be dropped altogether. This approximation is valid because the contribution of chloride to the resting membrane potential is insignificant in most nerve cells. In this case, the Goldman equation becomes

$$E_m = 58 \text{ mV } \log\left(\frac{[\text{K}^+]_o + b[\text{Na}^+]_o}{[\text{K}^+]_i + b[\text{Na}^+]_i}\right) \qquad (4\text{-}3)$$

This is the form typically encountered in neurophysiology. In nerve cells, the ratio of sodium to potassium permeability, b, is commonly about 0.02, although this value may vary somewhat from one type of cell to another. That is, p_K is about 50 times higher than p_{Na}. Thus, equation (4-3) tells us that E_m would be about -71 mV for a cell with $[\text{K}^+]_i = 125$ mM, $[\text{K}^+]_o = 5$ mM, $[\text{Na}^+]_i = 12$ mM, $[\text{Na}^+]_o = 120$ mM, and $b = 0.02$. What would E_m be for the same cell if b were 1.0 (that is, if $p_{Na} = p_K$) instead of 0.02?

The Goldman equation tells us quantitatively what we would expect qualitatively. If p_K is 50 times higher that p_{Na}, we would expect E_m to be nearer to E_K than to E_{Na}. Indeed, equation (4-3) yields $E_m = -71$ mV, which is much nearer to E_K (-80 mV) than to E_{Na} ($+58$ mV). The difference between E_m and E_K reflects the steady influx of sodium ions carrying positive charge into the cell and maintaining a depolarization from E_K.

The applicability of the Goldman equation to a real cell can be tested experimentally by varying the concentration of potassium in the ECF and measuring the resulting changes in membrane potential. If membrane potential were determined solely by the distribution of potassium ions across the cell membrane—that is, if the factor b in equation (4-3) were 0—we know that E_m would be determined by the potassium equilibrium potential. In this situation, a plot of measured membrane potential against log $[\text{K}^+]_o$ would yield a straight line with a slope of 58 mV per tenfold change in $[\text{K}^+]_o$. This straight line would merely be a plot of the E_m calculated from the Nernst equation at different values for external potassium concentration, and it is shown by the dashed line in Figure 4-4. Look, however, at the actual data from a real experiment in Figure 4-4. These data show the measured values of E_m of a nerve fiber observed at a number of different external potassium concentrations. The data do not follow the line expected from the Nernst equation, but instead fall along the solid line. That line was drawn according to the form of the Goldman equation given in equation (4-3), and this experiment demonstrates that the real value of membrane potential in the nerve fiber is determined jointly by potassium and sodium ions. Experiments of this type by Hodgkin and Katz in 1949 first demonstrated the role of sodium ions in the resting membrane potential of real cells.

Equation (4-3) is a reasonable approximation to equation (4-2) only if p_{Cl}/p_K is negligible. To determine if it is valid to ignore the

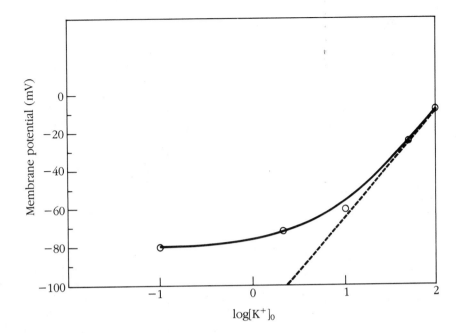

Figure 4-4
Experimentally determined relation between external potassium concentration and resting membrane potential of an axon in the spinal cord of the lamprey. The dashed line gives the potassium equilibrium potential calculated from the Nernst equation. The solid line shows the prediction from the Goldman equation with internal and external sodium and potassium concentrations appropriate for the lamprey nervous system.

contribution of chloride—that is, to use equation (4-3)—experiments like that summarized in Figure 4-4 can be performed in which the concentration of chloride in the ECF is varied rather than the concentration of potassium. When that was done on the type of nerve cell used in the experiment of Figure 4-4, it was found that a tenfold reduction of $[Cl^-]_o$ caused only a 2 mV change in the resting membrane potential. Thus, for that type of cell, membrane potential is relatively unaffected by chloride concentration, and equation (4-3) is valid. This is also true for other nerve cells. It is important to emphasize, however, that the membranes of other kinds of cells, such as muscle cells, have larger chloride permeability; therefore, the membrane potential of those cells would be more strongly dependent on external chloride concentration. This has been demonstrated experimentally for muscle cells by Hodgkin and Horowicz.

Ionic Steady State

The Goldman equation represents the actual situation in animal cells. The membrane potential of the cell takes on a steady value that reflects a fine balance between competing influences. It is important to keep in mind that neither sodium ions nor potassium ions are at equilibrium at that steady value of potential: sodium ions are continually leaking into the cell and potassium ions are continually leaking out. If this were allowed to continue, the concentration gradients for sodium and potassium would eventually run down and the membrane potential would decline to zero as the ionic gradients collapsed. It is like a flashlight that has been left on: the batteries slowly discharge.

To prevent the intracellular accumulation of sodium and loss of potassium, the cell must expend energy to restore the ionic gradients. Here again is an important role for the sodium pump. Metabolic energy stored in ATP is used to extrude the sodium that leaks in and to regain the potassium that was lost. In this way, the batteries are recharged using metabolic energy. Viewed in this light, we can see that the steady membrane potential of a cell represents chemical energy that has been converted into a different form and stored in the ionic gradients across the cell membrane. In Part II of this book, we will see how some cells, most notably the cells that make up the nervous system, are able to tap this stored energy to generate signals that can carry information and allow animals to move about and function in their environment.

The Chloride Pump

Because the resting membrane potential of a cell is not at either the sodium or potassium equilibrium potentials, there is a continuous net flux of sodium across the membrane. As we have just seen, metabolic energy must be expended in order to maintain the ionic gradients for sodium and potassium. What about chloride? The equilibrium potential for chloride given the internal and external concentrations shown in Table 1-1 would be about -80 mV, but the resting membrane potential is about -71 mV. Thus, we would expect that there would be a steady influx of chloride into the cell because of this imbalance between the electrical and concentration gradients for chloride. Eventually, this influx would raise the internal chloride concentration to the point where the new chloride equilibrium potential would be -71 mV, the same as the resting membrane potential. At that point the concentration gradient for chloride would be reduced sufficiently to come into balance with the resting membrane potential. We can calculate from the Nernst equation that chloride would have to rise to about 7.5 mM from its usual 5 mM in order for this new equilibrium state to be established.

In some cells, this does indeed appear to happen: chloride

reaches a new equilibrium governed by the resting membrane potential of the cell. (The cell would also gain the same small amount of potassium; because there is so much potassium inside, a change of a few millimolar in potassium concentration makes very little change in the potassium equilibrium potential, however.) In other cells, however, the chloride equilibrium potential remains different from the resting membrane potential, just as the sodium and potassium equilibrium potentials remain different from E_m. The only way this nonequilibrium condition can be maintained is by expending energy to keep the internal chloride constant—that is, there must also be a chloride pump similar in function to the sodium–potassium pump. In most cells, the chloride pump moves chloride ions out of the cell, so that the chloride equilibrium potential remains more negative than the resting membrane potential. In a few cases, however, an inwardly directed chloride pump has been discovered. Less is known about the molecular machinery of the chloride pump than that of the sodium–potassium pump. It is thought to involve an ATPase in some instances, so that the energy released by hydrolysis of ATP is the immediate driving energy for the pumping. In other cases, the pump may use energy stored in gradients of other ions to drive the movement of chloride.

Electrical Current and the Movement of Ions Across Membranes

An electrical current is the movement of charge through space. In a wire like that carrying electricity in your house, the electrical current is a flow of electrons; in a solution of ions, however, a flow of current is carried by movement of ions. That is, in a solution, the charges that move during an electrical current flow are the charges on the ions in solution. Thus, the movement of ions through space—such as from the outside of a cell to the inside of a cell— constitutes an electrical current, just as the movement of electrons through a wire constitutes an electrical current.

By thinking of ion flows as electrical currents, we can get a different perspective on the factors governing the steady-state membrane potential of cells. We have seen that at the steady-state value of membrane potential, there is a steady influx of sodium ions into the cell and a steady efflux of potassium ions out of the cell. This means that there is a steady electrical current, carried by sodium ions, flowing across the cell membrane in one direction and another current, carried by potassium ions, flowing across the membrane in the opposite direction. By convention, it is assumed that electrical current flows from the plus to the minus terminal of a battery; that is, we talk about currents in a wire as though the current is carried by positive charges. (As we all learn in introductory physics, this convention arose from a wrong guess by the early experimenters with electricity: electrons actually are the charge carriers in wires.) By extension, this convention means that the sodium current is an *inward* membrane current (the transfer of positive charge from the outside to the inside of the membrane), and the potassium current is an *outward* membrane current.

As we saw in our discussion of the Goldman equation above, a steady value of membrane potential will be achieved when the influx of sodium is exactly balanced by the efflux of potassium. In electrical terms, this means that in the steady state the sodium current, i_{Na}, is equal and opposite to the potassium current, i_K. In equation form, this can be written

$$i_K + i_{Na} = 0 \qquad (4\text{-}4)$$

Thus, at the steady state the net membrane current is zero. This makes electrical sense, if we keep in mind that the cell membrane can be treated as an electrical capacitor (Chapter 3). If the sum of i_{Na} and i_K were not zero, there would be a net flow of current across the membrane. Thus, there would be a movement of charge onto (or from) the membrane capacitor. Any such movement of charge would change the voltage across the capacitor (the membrane potential); that is, from the relation $q = CV$, if q changes and C remains constant, then V must of necessity change. Equation (4-4), then, is a requirement of the steady-state condition; if the equation is not true, the membrane potential cannot be at a steady level.

In cells in which there is an appreciable flow of chloride ions across the membrane, equation (4-4) must be expanded to include the chloride current, i_{Cl}:

$$i_K + i_{Na} + i_{Cl} = 0 \qquad (4\text{-}5)$$

Equation (4-5) is, in fact, the starting point in the derivation of the Goldman equation (see Appendix B). Note that because of the negative charge of chloride and because of the electrical convention for the direction of current flow, an outward movement of chloride ions is actually an inward membrane current.

Factors Affecting Ionic Current Across a Cell Membrane

What factors govern the amount of current carried across the membrane by a particular ion? We would expect that one important factor would be the difference between the equilibrium potential for the ion and the actual membrane potential. As an example, consider the movement of potassium ions across the membrane. We know that if $E_m = E_K$, there is a balance between the electrical and concentrational forces for potassium and there is no net movement of potassium across the membrane. In this situation, then, $i_K = 0$. As shown in Figure 4-2, if E_m does not equal E_K, the resulting imbalance in electrical and concentrational forces will drive a net movement of potassium across the membrane. The larger the difference between E_m and E_K, the larger the imbalance between the electrical and concentration gradients and the larger the net movement of potassium. Thus, i_K depends on $E_m - E_K$. This difference is called the **driving force** for membrane current carried by an ion.

We would also expect that the permeability of the membrane to an ion would be an important determinant of the amount of membrane current carried by that ion. If the permeability is high, the ionic current at a particular value of driving force will be higher than if the permeability were low. Thus, because p_K is much greater than p_{Na}, the potassium current resulting from a 10 mV difference between E_m and E_K will be much larger than the sodium current resulting from a 10 mV difference between E_m and E_{Na}. This is, in electrical terms, the reason that the steady-state membrane potential of a cell lies close to E_K rather than E_{Na}: in order for equation (4-4) to be obeyed, the driving force for sodium entry $(E_m - E_{Na})$ must be much greater than the driving force for potassium exit $(E_m - E_K)$.

Membrane Permeability Vs. Membrane Conductance

To place the discussion in the preceding section on more quantitative ground, it will be necessary to introduce a new concept that is closely related to membrane permeability: **membrane conductance**. The conductance of a membrane to an ion is an index of the ability of that ion to carry current across the membrane: the higher the conductance, the greater the ionic current for a given driving force. Conductance is analogous to the reciprocal of the resistance of an electrical circuit to current flow: the higher the resistance of a circuit, the lower the amount of current that flows in response to a particular voltage. This behavior of electrical circuits can be conveniently summarized by Ohm's law: $i = V/R$. Here, i is the current flowing through a resistor, R, in the presence of a voltage gradient, V. The equivalent form for the flow of an ionic current across a membrane is, using potassium as an example:

$$i_K = g_K(E_m - E_K) \tag{4-6}$$

where g_K is the conductance of the membrane to potassium ions. Similar equations can be written for sodium and chloride:

$$i_{Na} = g_{Na}(E_m - E_{Na}) \tag{4-7}$$

$$i_{Cl} = g_{Cl}(E_m - E_{Cl}) \tag{4-8}$$

Note that for the usual values of E_m (-71 mV), E_K (-80 mV), and E_{Na} ($+58$ mV), the potassium current is a positive number and the sodium current is a negative number, as required by the fact that the two currents flow in opposite directions across the membrane. By convention in neurophysiology, an outward membrane current (i_K, at steady-state E_m) is positive and an inward current (i_{Na}, at steady-state E_m) is negative.

The membrane conductance to an ion is closely related to the membrane permeability to that ion, but the two are not identical. The membrane current carried by a particular ion, and hence the membrane conductance to that ion, is proportional to the rate at which ions are crossing the membrane (that is, the ionic flux). That rate depends not only on the permeability of the membrane to the ion, but also on the number of available ions in the solution. As an example, imagine a cell membrane with many potassium channels (see Figure 4-5). The permeability of this membrane to potassium is thus high. If there are few potassium ions in solution, however, the chance is small that a K^+ will encounter a channel and cross the membrane. In this case, the potassium current will be low and the conductance of the membrane to K^+ will be low even though the permeability is high. On the other hand, if there are many potassium ions available to cross the membrane (Figure 4-5B), the chance that a K^+ will encounter a channel is high, and the rate of

(A) High permeability + few ions = low ionic current

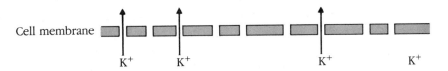

(B) High permeability + many ions = larger ionic current

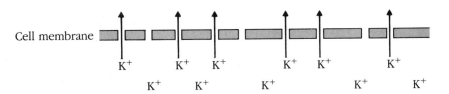

K^+ flow across the membrane will be high. The permeability remains fixed but the ionic conductance increases when more potassium ions are available. The point is that the potassium conductance of the membrane depends on the concentration of potassium at the membrane. For the most part, however, a change in permeability of a membrane to an ion produces a corresponding change in the conductance of the membrane to that ion. Thus, when we are dealing with changes in membrane conductance—as in the next chapter—we can treat a conductance change as a direct index of the underlying permeability change.

Figure 4-5
Illustration of the difference between permeability and conductance. (A) A cell membrane is highly permeable to potassium, but there is little potassium in solution. Therefore, the ionic current carried by potassium ions is small and the membrane conductance to potassium is small. (B) The same cell membrane in the presence of higher potassium concentration has a larger potassium conductance because the potassium current is larger. The permeability, however, is the same as in (A).

Summary

In real cells, the resting membrane potential is the point at which sodium influx is exactly balanced by potassium efflux. This point depends on the relative membrane permeabilities to sodium and potassium; in most cells p_K is much higher than p_{Na} and the balance is struck close to E_K. The Goldman equation gives the quantitative expression of the relation between membrane potential on the one hand and ion concentrations and permeabilities on the other. Because the steady-state membrane potential lies between the equilibrium potentials for sodium and potassium, there is a constant exchange of intracellular potassium for sodium. This would lead to progressive decline of the ionic gradients across the membrane if it were not for the action of the sodium–potassium pump. Thus, metabolic energy, in the form of ATP used by the pump, is required for the long-term maintenance of the sodium and potassium gradients. In the absence of chloride pumping, the chloride equilibrium potential will change to come into line with the value of membrane potential established by sodium and potassium. In some cells, however, a chloride pump maintains the internal chloride concentration in a nonequilibrium state, just as the sodium–potassium pump maintains internal sodium and potassium concentrations at nonequilibrium values.

The steady fluxes of potassium and sodium ions constitute electrical currents across the cell membrane, and at the steady-state E_m these currents cancel each other so that the net membrane current is zero. The membrane current carried by a particular ion is given by an ionic form of Ohm's law—that is, by the product of the driving force for that ion and the membrane conductance to that ion. The driving force is the difference between the actual value of membrane potential and the equilibrium potential for that ion. Conductance is a measure of the ability of the ion to carry electrical current across the membrane, and it is closely related to the membrane permeability to the ion.

CELLULAR PHYSIOLOGY OF NERVE CELLS

Part I of this book focused on general properties that are shared by all cells. Every cell must achieve osmotic balance, and all cells have an electrical membrane potential. Part II considers properties that are peculiar to particular kinds of cells: those capable of modulating their membrane potential in response to stimulation from their environment. These cells are called **excitable** cells because they can generate active electrical responses that can serve as signals or triggers for other events. The most notable examples of excitable cells are the cells of the nervous system, which are called neurons.

The nervous system must receive information from the external environment and from the internal environment (that is, from the body), transmit and analyze that information, and coordinate an appropriate action in response. These functions of the nervous system range from controlling digestion in the gut of a worm to the complex processing your brain is performing in reading these words. As far as we know, however, the underlying signals the nervous system uses to process information are the same in the worm and in the human brain. In all cases, the signals passed along in the nervous system are electrical signals, modulations of the membrane potential.

Part II will be concerned with those electrical signals, the currency of information in the nervous system. We will discuss how the signals arise, how a signal can be made to propagate throughout a nerve cell, and how the signals are passed along from one nerve cell to another. We will find that simple modifications of the scheme for the origin of the membrane potential presented in Chapter 4 can explain how neurons are capable of carrying out the vital signalling functions that are the task of the nervous system.

5

Generation of Nerve Action Potential

To set the stage for the discussion of the generation and transmission of signals in the nervous system, it will be useful to review the characteristics of those signals within a well-defined example: the **patellar reflex**. This is better known as the knee-jerk reflex and is familiar to most of us. It is a type of stretch reflex and is elicited by a sudden stretching of the quadriceps muscle at the front of the thigh. The neural circuitry that underlies the patellar reflex is diagrammed in Figure 5-1. A sharp blow to the patellar tendon, which connects the kneecap (patella) to the bones of the lower leg, pulls the kneecap down and stretches the quadriceps muscle. Special sensory neurons with endings in the muscle are activated by the stretch. The message that the muscle has been stretched travels from the muscle along the thin fibers of the sensory neurons, which terminate in the lower spinal cord. Activity in the sensory neurons is then passed along to other neurons in the spinal cord. Among these neurons are motor neurons, each of which sends a long, thin fiber back to the thigh to make contact with the quadriceps muscle. Activity in the motor neurons causes the muscle to contract.

Neurons are structurally complex cells, with long fibrous extensions that are specialized to receive and transmit information throughout the nervous system. This complexity can be appreciated by looking at the delicate structure of a motor neuron, shown in Figure 5-2. The cell body, or **soma**, of the motor neuron—where the nucleus resides—is only a small part of the cell. It is typically only about 20 to 30 μm in diameter in the case of motor neurons involved in the patellar reflex. The soma gives rise, how-

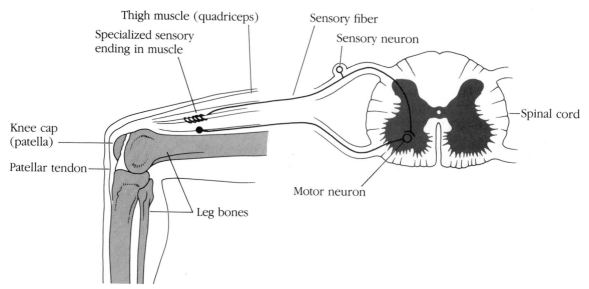

Thigh muscle (quadriceps)

Sensory fiber

Specialized sensory
ending in muscle

Sensory neuron

Knee cap
(patella)

Patellar tendon

Spinal cord

Motor neuron

Leg bones

Figure 5-1

Schematic representation of the
patellar reflex. The sensory
neuron is activated by stretching
the thigh muscle and in turn
activates motor neurons in the
spinal cord. Activity in the motor
neuron causes contraction of the
thigh muscle.

ever, to a tangle of profusely branching processes called **dendrites**, which might spread out for several millimeters within the spinal cord. The dendrites are specialized to receive signals passed along as the result of the activity of other neurons, such as the sensory neurons of the patellar reflex, and to funnel those signals along to the soma. The soma also gives rise to a thin fiber, called an **axon**, that is specialized to transmit activity from the neuron to other neurons, or in the case of the motor neuron, to the cells of the muscles. In some neurons, the axon may also branch profusely and carry signals to many parts of the nervous system. As shown in Figure 5-2B, the sensory neuron of the patellar reflex is structurally much simpler than the motor neuron. Its soma, which is located just outside the spinal cord in the **dorsal root ganglion**, gives rise to only a single process. This process bifurcates to form the axon that carries the signal generated by stretch from the muscle into the spinal cord.

In this chapter, we will be concerned only with the long-distance signal, the action potential, that carries the message along the axon of the sensory neuron from the muscle to the spinal cord and along the axon of the motor neuron from the spinal cord to the muscle. This chapter will concentrate on a descriptive introduction to the action potential and its mechanism; Chapter 6 will present in more advanced form the elegant physiological experiments that gave rise to the current view of that mechanism. In later chapters, we will discuss how the activity is transmitted from the sensory neuron to the motor neuron and how the muscle contracts in response to the activity of the motor neuron.

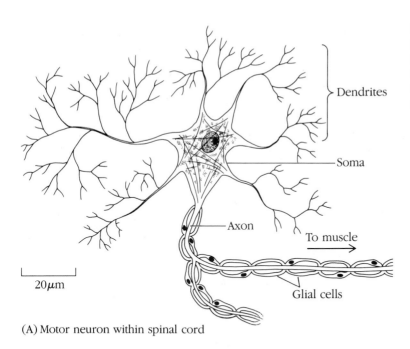

(A) Motor neuron within spinal cord

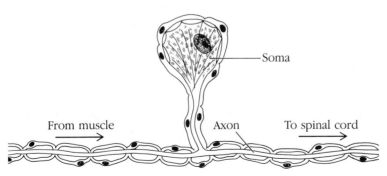

(B) Sensory neuron just outside spinal cord

Figure 5-2
Structures of single neurons involved in the patellar reflex. (A) Motor neuron within spinal cord. (B) Sensory neuron just outside spinal cord.

The Action Potential

Ionic Permeability and Membrane Potential

In Chapter 4, we saw that membrane potential is governed by the relative permeabilities of the permeant ions. If the sodium permeability of a cell is greater than the potassium permeability, the membrane potential will be closer to E_{Na} than to E_K. Conversely, if a cell has greater potassium than sodium permeability, E_m will lie closer to E_K. In all cases discussed so far, ionic permeability has

Figure 5-3
Example of an action potential in
a neuron. (A) Experimental
arrangement for recording the
membrane potential of a nerve
cell fiber. (B) Resting membrane
potential and an action potential
recorded via a voltage-sensing
probe inside the sensory neuron
of the patellar reflex loop.
(C) Actual intracellular recordings
from a single stretch-receptor
sensory neuron during stretch of
the muscle. The lower trace
shows a single action potential on
an expanded time scale to show
its waveform in more detail.
[Intracellular records of action
potentials in part C were traced
from unpublished data kindly
provided by R. Rose, M. Sedivec,
and L. M. Mendell of the State
University of New York, Stony
Brook.]

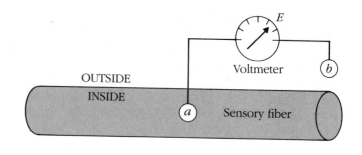

(A)

been treated as a fixed characteristic of the cell membrane. In this
chapter, however, we will see that the ionic permeability of the
plasma membrane of excitable cells can vary. Specifically, we will
see that a transient, dramatic increase in sodium permeability
underlies the generation of the basic signal of the nervous system,
the action potential.

Measuring the Long-Distance Signal in Neurons

What kind of signal carries the message along the sensory neuron
in the patellar reflex? As mentioned earlier, the signals in the
nervous system are electrical signals, and to monitor these signals it
is necessary to measure the changes in electrical potential associated with the activation of the reflex. This can be done by placing an
ultrafine probe inside the sensory axon and using this probe to
measure the electrical membrane potential of the neuron. A diagram illustrating this kind of experiment is shown in Figure 5- 3. A
voltmeter has been connected to measure the voltage difference
between point *a*, at the tip of the ultrafine probe, and point *b*, a
reference point in the ECF. As shown in Figure 5-3B, when the
probe is outside the sensory axon, both the probe and the reference point are in the ECF, and there is no voltage difference measured by the voltmeter. When the probe is inserted into the sensory
fiber, however, we measure the voltage difference between the
inside and outside of the neuron, the membrane potential. In
neurophysiology, this process of measuring the transmembrane
voltage of a cell is called intracellular recording, and the ultrafine
probe is called an intracellular electrode. As expected from the
discussion in Chapter 4, the membrane potential of the sensory
fiber is about −70 mV.

When the muscle is stretched (Figure 5-3B), something dramatic

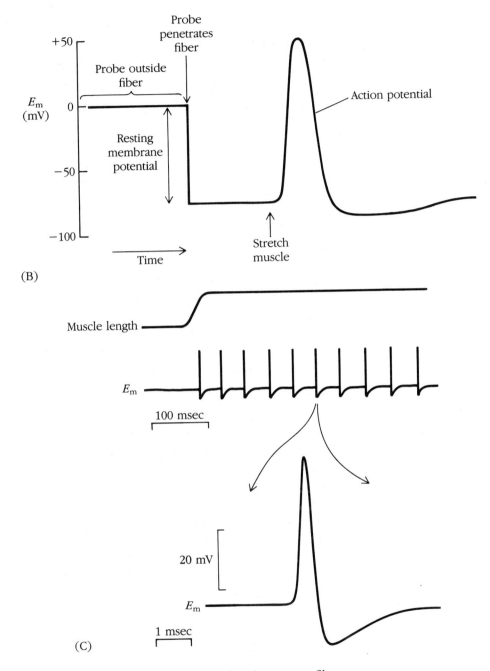

(B)

(C)

happens to the measured membrane potential in the sensory fiber. After a small delay, there is a sudden jump in membrane potential in which the potential transiently moves in a positive direction (a depolarization) and actually reverses in sign for a brief period. When the potential returns toward its resting value, it may transiently become more negative than its normal resting value. The transient jump in potential is called an action potential, which is the long-distance signal of the nervous system. If the stretch is sufficiently strong, it might elicit a series of several action potentials, each with the same shape and amplitude. Figure 5-3C shows examples of action potentials recorded via an intracellular electrode inside a sensory axon during stretch of a muscle.

Characteristics of the Action Potential

The action potential has several important characteristics that will be explained in terms of the underlying ionic permeability changes. These include the following:

1. Action potentials are triggered by depolarization.
The stimulus that initiates an action potential in a neuron is a reduction in the membrane potential, that is, a depolarization of the membrane. Normally, the depolarization is produced by some external stimulus, such as the stretching of the muscle in the case of the sensory neuron in the patellar reflex, or by the action of another neuron, as in the transmission of excitation from the sensory neuron to the motor neuron in the patellar reflex.

2. A threshold level of depolarization must be reached in order to trigger an action potential. A small depolarization from the normal resting membrane potential will not produce an action potential. In most neurons, it is necessary to depolarize the membrane by about 10 to 20 mV in order to trigger an action potential; therefore, if a neuron has a resting membrane potential of about −70 mV, we would probably have to reduce the membrane potential to about −60 to −50 mV in order to trigger an action potential.

3. Action potentials are all-or-none events. This means that once a stimulus is strong enough to reach threshold, the amplitude of the action potential is independent of the strength of the stimulus. In other words, the amplitude and form of the action potential are fixed: the event either occurs all the way (if a depolarization is above threshold) or it doesn't occur at all (if the depolarization is below threshold). In this manner, the triggering of an action potential is like the firing of a gun: the speed with which the bullet leaves the barrel is independent of whether the trigger was pulled softly or hard.

4. An action potential propagates without decrement throughout a neuron, but it does so at a speed much slower than the rate at which electricity propagates through a wire. Figure 5-3C shows intracellular recordings of the action potentials triggered in a sensory neuron by stretching the quadriceps muscle. If we recorded simultaneously from the sensory fiber near the muscle and near the spinal cord, we would find that the action potential at the two locations has the same amplitude and form. Thus, as the signal travels from the muscle—where it originated—to the spinal cord, its amplitude remains unchanged. However, there would be a significant delay, almost 0.1 sec, between the appearance of the action potential near the muscle and its arrival at the spinal cord. The conduction speed in a typical mammalian nerve fiber is about 10 to 20 m/sec, although speeds as high as 100 m/sec have been observed.

5. At the peak of the action potential, the membrane potential reverses sign, becoming inside positive. As shown in Figure 5-3, the membrane potential during an action potential transiently overshoots zero, so that the inside of the cell becomes positive with respect to the ECF for a brief time. This is called the overshoot of the action potential. When the action potential repolarizes toward the normal resting membrane potential, it transiently becomes more negative than normal. This is called the undershoot of the action potential.

6. After a neuron fires an action potential, there is a brief period, called the absolute refractory period, during which it is impossible to make the neuron fire another action potential. The absolute refractory period varies a bit from one neuron to another, but it usually lasts about 1 msec. This sets a maximum limit of about 1000 action potentials per second on the rate at which a neuron can fire.

The goal of the remainder of this chapter is to explain all of these characteristics of the nerve action potential in terms of changes in the ionic permeability of the cell membrane and the resulting movements of ions.

Initiation and Propagation of Action Potentials

Some of the fundamental properties of action potentials can be studied experimentally using an apparatus like that diagrammed in Figure 5-4A. Imagine that a long section of a single axon is removed and arranged in the apparatus so that intracellular probes can be placed inside the fiber at two points, *a* and *b*, which are 10 cm apart. The probe at *a* is set up to pass positive or negative charge into the fiber and to record the resulting change in membrane potential, while the probe at *b* records membrane potential only. The effect of injecting negative charge at a constant rate at *a* is shown in Figure 5-4B. The extra negative charges make the interior of the fiber more negative, and the membrane potential increases; that is, the membrane is hyperpolarized. At the same time, the probe at *b* records no change in membrane potential at all. This is because the plasma membrane is leaky to charge: it is not a very good electrical insulator. In Chapter 3, we discussed the cell membrane as an electrical capacitor. In addition, the membrane behaves like an electrical resistor; that is, there is a direct path through which ionic current may flow across the membrane. As we saw in Chapter 4, that current path is through the ion channels that are inserted into the lipid bilayer of the plasma membrane. Thus, the charges injected at *a* do not travel very far down the fiber before leaking out of the cell across the plasma membrane. None of the charges reaches *b*, and so there is no change in membrane potential at *b*. When we stop injecting negative charges at *a*, all the injected charge leaks out of the cell, and the membrane potential returns to its normal resting value.

Another way of looking at the situation in Figure 5-4B is in terms of the flow of electrical current. The negative charges injected into the cell at a constant rate constitute an electrical current originating from the experimental apparatus. The return path for the current to the apparatus lies in the ECF, so that in order to complete the

Figure 5-4

Demonstration of action potential generation and propagation in a segment of a nerve fiber. (A) Apparatus for recording electrical activity of a segment of a sensory nerve fiber. The probes at points *a* and *b* allow recording of membrane potential, and the probe at *a* also allows charges to be injected into the fiber. (B) Injecting negative charges at *a* causes a hyperpolarization at *a*. All injected charges leak out across the membrane before reaching *b*, and there is no change in membrane potential at *b*. (C) Injection of a small amount of positive charge produces a depolarization at *a* that again does not reach *b*. (D) If a stronger depolarization is induced at *a*, an action potential is generated. The action potential propagates without decrement along the fiber and is recorded at full amplitude at *b*.

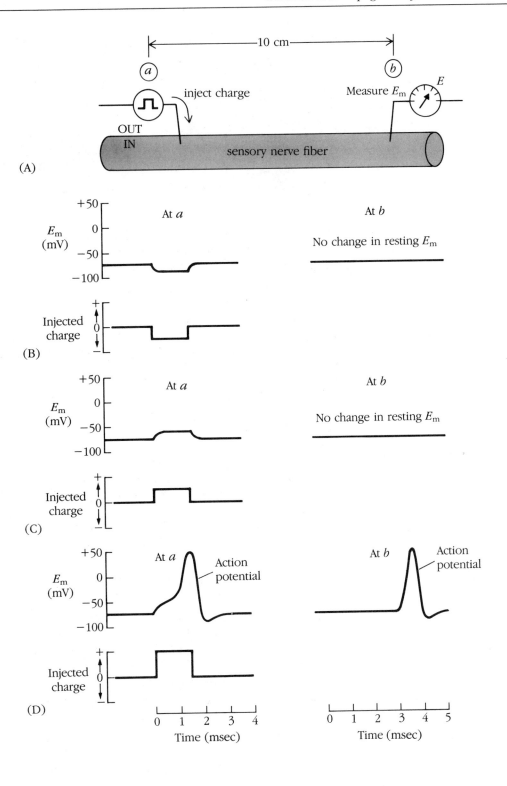

Figure 5-5

Schematic representation of the decay of injected current in an axon with distance from the site of current injection.

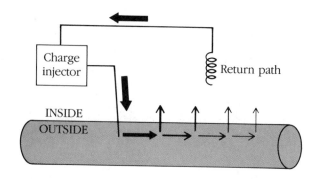

circuit the current must exit across the plasma membrane. Two paths are available for the current at the point where it is injected: it can flow across the membrane immediately, or it can move down the axon to flow out through a more distant segment of axon membrane. This situation is illustrated in Figure 5-5. The injected current will thus divide, some taking one path and some the other. The proportion of current taking each path depends on the relative resistances of the two paths: more current will flow down the path with less resistance. With each increment in distance along the axon, that fraction of the injected current that flowed down the axon again faces two paths; it can continue down the interior of the axon or it can cross the membrane at that point. The current will again divide, and some fraction of the fraction of the injected current will continue down the nerve fiber. This will continue until all the injected current has crossed the membrane; at that point, the injected current will not influence the membrane potential because there will be no remaining injected current. Thus, the change in membrane potential produced by current injection (Figure 5-4A) decays with distance from the injection site. The greatest effect occurs at the injection site, and there is progressively less effect as injected current is progressively lost across the plasma membrane. The cell membrane is not a particularly good insulator (it has a low resistance to current flow compared, for example, with the insulator surrounding the electrical wires in your house), and the ICF inside the axon is not a particularly good conductor (its resistance to current flow is high compared with that of a copper wire). This set of circumstances favors the rapid decay of injected current with distance. In real axons, the hyperpolarization produced by current injected at a point decays by about 95% within 1 to 2 mm of the injection site.

Let's return now to the experiment shown in Figure 5-4. The effect of injecting positive charges into the axon is shown in Figure

5-4C. If the number of positive charges injected is small, the effect is simply the reverse of the effect of injecting negative charges; the membrane depolarizes while the charges are injected, but the effect does not reach *b*. When charge injection ceases, the extra positive charges leak out of the fiber, and membrane potential returns to normal. If the rate of injection of positive charge is increased, as in Figure 5-4D, the depolarization is larger. If the depolarization is sufficiently large, an all-or-none action potential, like that recorded when the muscle was stretched (Figure 5-3), is triggered at *a*. The probe at *b* records a replica of the action potential at *a*, except that there is a time delay between the occurrence of the action potential at *a* and its arrival at *b*. Thus, action potentials are triggered by depolarization, not by hyperpolarization (characteristic 1, above), the depolarization must be large enough to exceed a threshold value (characteristic 2), and the action potential travels without decrement throughout the nerve fiber (characteristic 4). What ionic properties of the neuron membrane can explain these properties?

Changes in Relative Sodium Permeability During an Action Potential

The key to understanding the origin of the action potential lies in the discussion in Chapter 4 of the factors that influence the steady-state membrane potential of a cell. Recall that the resting E_m for a neuron will lie somewhere between E_K and E_{Na}. According to the Goldman equation, the exact point at which it lies will be determined by the ratio p_{Na}/p_K. As we saw in Chapter 4, p_{Na}/p_K of a resting neuron is about 0.02, and E_m is near E_K.

What would happen to E_m if sodium permeability suddenly increased dramatically? The effect of such an increase in p_{Na} is diagrammed in Figure 5-6. In the example, p_{Na} undergoes an abrupt thousandfold increase, so that $p_{Na}/p_K = 20$ instead of 0.02. According to the Goldman equation, E_m would then swing from about -70 mV to about $+50$ mV, near E_{Na}. When p_{Na}/p_K returns to 0.02, E_m will return to its usual value near E_K. Note that the swing in membrane potential in Figure 5-6 reproduces qualitatively the change in potential during an action potential. Indeed, it is a transient increase in sodium permeability, as in Figure 5-6, that is responsible for the swing in membrane polarization from near E_K to near E_{Na} and back during an action potential.

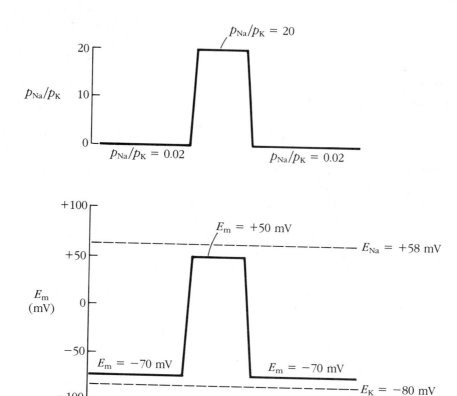

$p_{Na}/p_K = 20$

$p_{Na}/p_K = 0.02$

$p_{Na}/p_K = 0.02$

$E_m = +50$ mV

$E_{Na} = +58$ mV

$E_m = -70$ mV

$E_m = -70$ mV

$E_K = -80$ mV

Figure 5-6

The relation between relative sodium permeability and membrane potential. When the ratio of sodium to potassium permeability (upper trace) is changed, the position of E_m relative to E_K and E_{Na} changes accordingly.

Voltage-Dependent Sodium Channels of the Neuron Membrane

Recall that ions must cross the membrane through transmembrane pores or channels. A dramatic increase in sodium permeability like that shown in Figure 5-6 requires a dramatic increase in the number of membrane channels that allow sodium ions to enter the cell. Thus, the resting p_{Na} of the membrane of an excitable cell is only a small fraction of what it could be because most membrane sodium channels are closed at rest. What stimulus causes these hidden channels to open and produces the positive swing of E_m during an action potential? It turns out that the conducting state of sodium channels of excitable cells depends on membrane potential. When E_m is at the usual resting level or more negative, these sodium channels are closed, Na$^+$ cannot flow through them, and p_{Na} is low. These channels open, however, when the membrane is depolarized. The stimulus for opening of the voltage-dependent sodium channels of excitable cells is a reduction of the membrane potential.

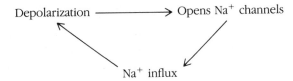

Figure 5-7
Explosive cycle leading to
depolarizing phase of an action
potential.

Because the voltage-dependent sodium channels respond to depolarization, the response of the membrane to depolarization is regenerative, or explosive. This is illustrated in Figure 5-7. When the membrane is depolarized, p_{Na} increases, allowing sodium ions to carry positive charge into the cell. This depolarizes the cell further, causing a greater increase in p_{Na} and more depolarization. Such a process is inherently explosive and tends to continue until all sodium channels are open and the membrane potential has been driven up to near E_{Na}. This explains the all-or-none behavior of the nerve action potential: once triggered, the process tends to run to completion.

Why should there be a threshold level of depolarization? Under the scheme discussed above, it might seem that any small depolarization would set the action potential off. However, in considering the effect of a depolarization, we must take into account the *total* current that flows across the membrane in response to the depolarization, not just the current carried by sodium ions. Recall that, at the resting E_m, p_K is very much greater than p_{Na}; therefore, flow of K^+ out of the cell can counteract the influx of Na^+ even if p_{Na} is moderately increased by a depolarization. Thus, for a moderate depolarization, the efflux of potassium might be larger than the influx of sodium, resulting in a net outward membrane current that keeps the membrane potential from depolarizing further and prevents the explosive cycle underlying the action potential. In order for the explosive process to be set in motion and an action potential to be generated, a depolarization must produce a *net inward* membrane current, which will in turn produce a further depolarization. A depolarization that produces an action potential must be sufficiently large to open quite a few sodium channels in order to overcome the efflux of potassium ions resulting from the depolarization. The **threshold potential** will lie at that value of E_m where the influx of Na^+ exactly balances the efflux of K^+; any further depolarization will allow Na^+ influx to dominate, resulting in an explosive action potential.

Factors that influence the actual value of the threshold potential for a particular neuron include the density of voltage-sensitive sodium channels in the plasma membrane and the strength of the connection between depolarization and opening of those channels. Thus, if voltage-sensitive sodium channels are densely packed in the membrane, opening only a small fraction of them will produce a sizeable inward sodium current, and we would expect that the threshold depolarization would be smaller than if the channels were sparse. Often, the density of voltage-sensitive sodium channels is highest just at the point (called the **initial segment**) where a neuron's axon leaves the cell body; this results in that portion of the cell having the lowest threshold for action potential generation. Another important factor in determining the threshold is the steepness of the relation between depolarization and sodium channel opening. In some cases the sodium channels have "hair triggers," and only a small depolarization from the resting E_m is required to open large numbers of channels. In such cases we would expect the threshold to be close to the resting membrane potential. In other neurons, larger depolarizations are necessary to open appreciable numbers of sodium channels, and the threshold is further from resting E_m.

Repolarization

What causes E_m to return to rest again following the regenerative depolarization during an action potential? There are two important factors: (1) the depolarization-induced increase in p_{Na} is transient; and (2) there is a delayed, voltage-dependent increase in p_K. These will be discussed in turn below.

The effect of depolarization on the voltage-dependent sodium channels is twofold. These effects can be summarized by the diagram in Figure 5-8, which illustrates the behavior of a single voltage-sensitive sodium channel in response to a depolarization. The channel acts as though the flow of Na^+ is controlled by two independent gates. One gate, called the *m* gate, is closed when E_m is equal to or more negative than the usual resting potential. This gate thus prevents Na^+ from entering the channel at the resting potential. The other gate, called the *h* gate, is open at the usual resting E_m. Both gates respond to depolarization, but with different speeds and in opposite directions. The *m* gate opens rapidly in response to depolarization; the *h* gate closes in response to depolarization, but does so slowly. Thus, immediately after a depolarization, the *m* gate is open, allowing Na^+ to enter the cell, but the *h* gate has not had time to respond to the depolarization and is

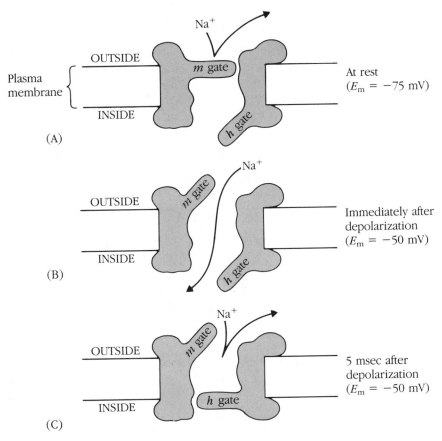

Na$^+$

OUTSIDE

m gate

Plasma membrane {

INSIDE

h gate

(A)

At rest
($E_m = -75$ mV)

Na$^+$

OUTSIDE

m gate

INSIDE

h gate

(B)

Immediately after depolarization
($E_m = -50$ mV)

Na$^+$

OUTSIDE

m gate

INSIDE

h gate

(C)

5 msec after depolarization
($E_m = -50$ mV)

thus still open. A little while later (about a millisecond or two), the *m* gate is still open, but the *h* gate has responded by closing, and the channel is again closed. The result of this behavior is that p_{Na} first increases in response to a depolarization, then declines again even if the depolarization were maintained in some way. This delayed decline in sodium permeability upon depolarization is called **sodium inactivation**. As shown in Figure 5-6, this return of p_{Na} to its resting level would alone be sufficient to bring E_m back to rest.

In addition to the voltage-sensitive sodium channels, there are voltage-sensitive potassium channels in the membranes of excitable cells. These channels are also closed at the normal resting membrane potential. Like the sodium channel *m* gates, the gates on the potassium channels open upon depolarization, so that the channel begins to conduct K$^+$ when the membrane potential is reduced. However, the gates of these potassium channels, which

Figure 5-8
Schematic representation of the behavior of a single voltage-sensitive sodium channel in the plasma membrane of a neuron.
(A) The state of the channel at normal resting membrane potential. (B) Upon depolarization, the *m* gate opens rapidly and sodium ions are free to move through the channel.
(C) After a brief delay, the *h* gate closes, returning the channel to a nonconducting state.

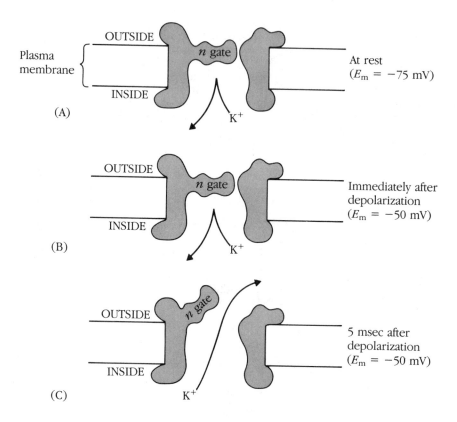

OUTSIDE

Plasma membrane

n gate

At rest
($E_m = -75$ mV)

INSIDE

K^+

(A)

OUTSIDE

n gate

Immediately after
depolarization
($E_m = -50$ mV)

INSIDE

K^+

(B)

OUTSIDE

n gate

5 msec after
depolarization
($E_m = -50$ mV)

INSIDE

K^+

(C)

Figure 5-9

The behavior of a single voltage-sensitive potassium channel in the plasma membrane of a neuron. (A) At the normal resting membrane potential, the channel is closed. (B) Immediately after a depolarization, the channel remains closed. (C) With a delay, the *n* gate opens, allowing potassium ions to cross the membrane through the channel. The channel remains open as long as depolarization is maintained.

are called *n* gates, respond slowly to depolarization, so that p_K increases with a delay following a depolarization. The characteristic behavior of a single voltage-sensitive potassium channel is shown in Figure 5-9. Unlike the sodium channel, there is no gate on the potassium channel that closes upon depolarization; the channel remains open as long as the depolarization is maintained and closes only when membrane potential returns to its normal resting value.

These voltage-sensitive potassium channels respond to the depolarizing phase of the action potential and open at about the time sodium permeability returns to its normal low value as *h* gates close. Therefore, the repolarizing phase of the action potential is produced by the simultaneous decline of p_{Na} to its resting level and increase of p_K to a higher than normal level. Note that during this time, p_{Na}/p_K is actually smaller than its usual resting value. This explains the undershoot of membrane potential below its resting value at the end of an action potential: E_m approaches closer to E_K because p_K is still higher than usual while p_{Na} has returned to its resting state. Membrane potential returns to rest as the slow n gates have time to respond to the repolarization by closing and returning p_K to its normal value.

The sequence of changes during an action potential is summarized in Figure 5-10, and characteristics of the various gates are summarized in Table 5-1. An action potential would be generated in the sensory neuron of the patellar reflex in the following way. Stretch of the muscle induces depolarization of the specialized sensory endings of the sensory neuron (probably by increasing the relative sodium permeability). This depolarization causes the m gates of voltage-sensitive sodium channels in the neuron membrane to open, setting in motion a regenerative increase in p_{Na}, which drives E_m up near E_{Na}. With a delay, h gates respond to the depolarization by closing and potassium channel n gates respond by opening. The combination of these delayed gating events drives E_m back down near E_K and actually below the usual resting E_m. Again with a delay, the repolarization causes the h gates to open and the n gates to close, and the membrane returns to its resting state, ready to respond to any new depolarizing stimulus.

The scheme for the ionic changes underlying the nerve action potential was worked out in a series of elegant electrical experiments by A. L. Hodgkin and A. F. Huxley of Cambridge University. Chapter 6 describes those experiments and presents a quantitative version of the scheme shown in Figure 5-10.

The Refractory Period

The existence of a refractory period would be expected from the gating scheme summarized in Figure 5-10. When the h gates of the voltage-sensitive sodium channels are closed (states C and D in Figure 5-10), the channels cannot conduct Na^+ no matter what the state of the m gate might be. When the membrane is in this condition, no amount of depolarization can cause the cell to fire an action potential; the h gates would simply remain closed, preventing the influx of Na^+ necessary to trigger the regenerative explosion. Only when enough time has passed for a significant number of h gates to reopen will the neuron be capable of producing another action potential.

Table 5-1 Summary of responses of voltage-sensitive sodium and potassium channels to depolarization.

Type of channel	Gate	Response to depolarization	Speed of response
Sodium	*m* gate	Opens	Fast
Sodium	*h* gate	Closes	Slow
Potassium	*n* gate	Opens	Slow

Propagation of an Action Potential Along a Nerve Fiber

We can now see how an action potential arises as a result of a depolarizing stimulus, such as the muscle stretch in the case of the sensory neuron of Figure 5-1. How does that action potential travel from the ending in the muscle along the long, thin sensory fiber to the spinal cord? The answer to this question is inherent in the scheme for generation of the action potential just presented. As we've seen, the stimulus for an action potential is a depolarization of greater than about 10 to 20 mV from the normal resting level of membrane potential. The action potential itself is a depolarization much in excess of this threshold level. Thus, once an action potential occurs at one end of a neuron, the strong depolarization will bring the neighboring region of the cell above threshold, setting up a regenerative depolarization in that region. This will in turn bring the next region above threshold, and so on. The action potential can be thought of as a self-propagating wave of depolarization sweeping along the nerve fiber. When the sequence of permeability changes summarized in Figure 5-10 occurs in one region

Figure 5-10
The state of voltage-sensitive sodium and potassium channels at various times during an action potential in a neuron. (A) At rest, neither channel is in a conducting state. (B) During the depolarizing phase of the action potential, sodium channels open, but potassium channels have not had time to respond to the depolarization. (C) During the repolarizing phase, sodium permeability begins to return to its resting level as *h* gates have sufficient time to respond to the depolarizing phase. At about the same time, potassium channels respond to the depolarization and open. (D) During the undershoot, sodium permeability has returned to its usual low level; potassium permeability, however, remains elevated because *n* gates respond slowly to the repolarization of the membrane. The resting state of the membrane is restored as *h* gates and *n* gates have time to respond to the repolarization.

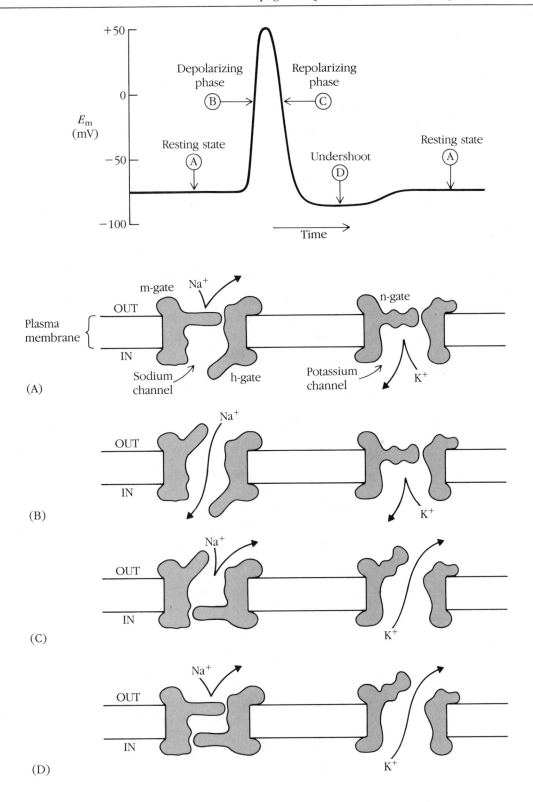

of a nerve membrane, it guarantees that the same gating events will be repeated in neighboring segments of membrane. In this manner, the cyclical change in membrane permeability, and the resulting action potential, chews its way along the nerve fiber from one end to the other, as each segment of axon membrane responds in turn to the depolarization of the preceding segment. This behavior is analogous to that of a lighted fuse, in which the heat generated in one segment of the fuse serves to ignite the neighboring segment.

A more formal description of propagation can be achieved by considering the electrical currents that flow along a nerve fiber during an action potential. Imagine that we freeze an action potential in time while it is traveling down an axon, as shown in Figure 5-11A. We have seen that at the peak of the action potential, there is an inward flow of current, carried by sodium ions. This is shown by the inward arrows at the point labeled 1 in Figure 5-11A. The region of axon occupied by the action potential will be depolarized with respect to more distant parts of the axon, like those labeled 2 and 3. This difference in electrical potential means that there will be a flow of depolarizing current leaving the depolarized region and flowing along the inside of the nerve fiber; that is, positive charges will move out from the region of depolarization. In the discussion of the response to injected current in an axon (see Figures 5-4 and 5-5), we saw that a voltage change produced by injected current decayed with distance from the point of injection. Similarly, the depolarization produced by the influx of sodium ions during an action potential will decay with distance from the region of membrane undergoing the action potential. This decay of depolarization with distance reflects the progressive leakage of the depolarizing current across the membrane, which occurs because the membrane is a leaky insulator. Figure 5-11B illustrates the profile of membrane potential that might be observed along the length of the axon at the instant the action potential at point 1 reaches its peak. Note that there is a region of axon over which the depolarization, although decaying, is still above the threshold for generating an action potential in that part of the membrane. Thus, if we "unfreeze" time and allow events to move along, the region that is above threshold will generate its own action potential. This process will continue as the action potential sweeps along the axon, bringing each sucessive segment of axon above threshold as it goes.

The flow of depolarizing current from the region undergoing an

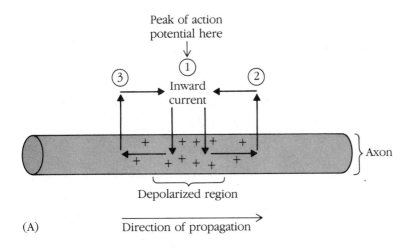

Peak of action
potential here

Inward
current

Axon

Depolarized region

(A)

Direction of propagation

Figure 5-11
The decay of depolarization with
distance from the peak of the
action potential at a particular
instant during the propagation of
the action potential from left to
right along the axon.

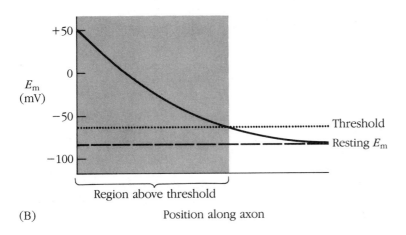

E_m
(mV)

+50

0

−50

−100

Threshold

Resting E_m

Region above threshold

(B)

Position along axon

action potential is symmetrical in both directions along the axon, as
shown in Figure 5-11A. Thus, current flows from point 1 to both
point 2 and to point 3 in the figure. Nevertheless, the action potential in an axon typically moves in only one direction. That is because the region the action potential has just traversed, like point 3,
is in the refractory period phase of the action potential cycle and
is thus incapable of responding to the depolarization originating
from the action potential at point 1. Of course, if a neurophysiologist comes along with an artificial situation, like that shown in
Figure 5-4, and stimulates an action potential in the middle of a
nerve fiber, that action potential will propagate in both directions
along the fiber. The normal direction of propagation in an axon—
the direction taken by normally occurring action potentials—is

called the **orthodromic** direction; an abnormal action potential propagating in the opposite direction is called an **antidromic** action potential.

Factors Affecting the Speed of Action Potential Propagation

The speed with which an action potential moves down an axon varies considerably from one axon to another; the range is from about 0.1m/sec to 100 m/sec. What characteristics of an axon are important in determining the action potential propagation velocity? Examine Figure 5-11B again. Clearly, if the rate at which the depolarization falls off with distance is less, the region of axon brought above threshold by an action potential at point 1 will be larger. If the region above threshold is larger, then an action potential at a particular location will set up a new action potential at a greater distance down the axon and the rate at which the action potential moves down the fiber will be greater. The rate of voltage decrease with distance will in turn depend on the relative resistance to current flow of the plasma membrane and the intracellular path down the axon. Recall from the discussion of the response of an axon to injection of current (see Figure 5-5) that there are always two paths that current flowing down the inside of axon at a particular point can take: it can continue down the interior of the fiber or cross the membrane at that point. We said that the portion of the current taking each path depends on the relative resistances of the two paths. If the resistance of the membrane could be made higher or if the resistance of the path down the inside of the axon could be made lower, the path down the axon would be favored and a larger portion of the current would continue along the inside. In this situation, the depolarization resulting from an action potential would decay less rapidly along the axon; therefore, the rate of propagation would increase.

Thus, two strategies can be employed to increase the speed of action potential propagation: increase the electrical resistance of the plasma membrane to current flow, or decrease the resistance of the longitudinal path down the inside of the fiber. Both strategies have been adopted in nature. Among invertebrate animals, the strategy has been to decrease the longitudinal resistance of the axon interior. This can be accomplished by increasing the diameter of the axon. When a fiber is fatter, it offers a larger cross-sectional area to the internal flow of current; the effective resistance of this larger area is less because the current has many parallel paths to choose from if it is to continue down the interior of the axon. For the same reason, the electric power company uses large-diameter copper wires for the cables leaving a power-generating station; these cables must carry massive currents and thus must have low resistance to current flow to avoid burning up. Some invertebrate axons are the neuronal equivalent of these power cables: axons up to 1 mm in diameter are found in some invertebrates. As expected, these giant axons are the fastest-conducting nerve fibers of the invertebrate world.

Among vertebrate animals, there is also some variation in the size of axons, which range from less than 1 μm in diameter to about 10 to 20 μm in diameter. Thus, even the largest axons in a human nerve do not begin to rival the size of the giant axons of invertebrates. Nevertheless, the fastest-conducting vertebrate axons are actually faster than the giant invertebrate axons. Vertebrate animals have adopted the strategy of increasing the membrane resistance to current as well as increasing internal diameter. This has been accomplished by wrapping the axon with extra layers of insulating cell membrane: in order to reach the exterior, electrical current must flow not only through the resistance of the axon membrane, but also through the cascaded resistance of the tightly wrapped layers of extra membrane. Figure 5-12A shows a schematic cross section of a vertebrate axon wrapped in this way. The cell that provides the spiral of insulating membrane surrounding the axon is a type of **glial cell**, a non-neuronal supporting cell of the nervous system that provides a sustaining mesh in which the neurons are imbedded.

The insulating sheath around the axon is called **myelin**. By increasing the resistance of the path across the membrane, the myelin sheath forces a larger portion of the current flowing as the result of voltage change to move down the interior of the fiber. This increases the spatial spread of a depolarization along the axon and increases the rate at which an action potential propagates. In order to set up a new action potential at a distant point along the axon, however, the influx of sodium ions carrying the depolarizing current during the initiation of the action potential must have access to the axon membrane. To provide that access, there are periodic breaks in the myelin sheath, called **nodes of Ranvier**, at regular intervals along the length of the axon. This is diagrammed in Figure 5-12B. Thus, the depolarization resulting from an action potential at

Figure 5-12
Propagation of an action potential along a myelinated nerve fiber. (A) Cross section of a myelinated axon, showing the spiral wrapping of the glial cell membrane around the axon. (B) The depolarization from an action potential at one node spreads far along the interior of the fiber because the insulating myelin prevents the leakage of current across the plasma membrane.

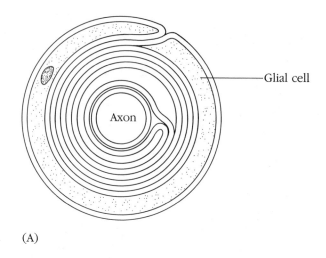

(A)

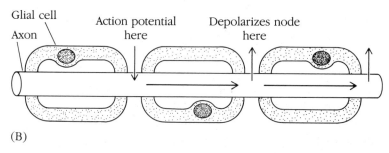

(B)

one node of Ranvier spreads along the interior of the fiber to the next node, where it sets up a new action potential. The action potential leaps along the axon, jumping from one node to the next. This form of action potential conduction is called **saltatory conduction**, and it produces a dramatic improvement in the speed with which a thin axon can conduct an action potential along its length.

The myelin sheath also has an effect on the behavior of the axon as an electrical capacitor. Recall from Chapter 3 that the cell membrane can be viewed as an insulating barrier separating two conducting compartments (the ICF and ECF). Thus, the cell membrane forms a capacitor. The capacitance, or charge-storing ability, of a capacitor is inversely related to the distance between the conducting plates: the smaller the distance, the greater the number of charges that can be stored on the capacitor in the presence of a particular voltage gradient. Thus, when the myelin sheath wrapped around an axon increases the distance between the conducting ECF and ICF, the effective capacitance of the membrane decreases. This means that a smaller number of charges needs be added to the inside of the membrane in order to reach a particular level of depolarization. (If it is unclear why this is true, review the calculation in Chapter 3 of the number of charges on a membrane at a particular voltage.) An electrical current is defined as the rate of charge movement—that is, number of charges per second. In the presence of a particular depolarizing current, then, a given level of voltage will be reached faster on a small capacitor than on a large capacitor. Because the myelin makes the membrane capacitance smaller, a depolarization will spread faster, as well as farther, in the presence of myelin.

Summary

The basic long-distance signal of the nervous system is a self-propagating depolarization called the action potential. The action potential arises because of a sequence of voltage-dependent changes in the ionic permeability of the neuron membrane. This voltage-dependent behavior of the membrane is due to gated sodium and potassium channels. The conducting state of the sodium channels is controlled by m gates, which are closed at the usual resting E_m and open rapidly upon depolarization, and by h gates, which are open at the usual resting E_m and close slowly upon depolarization. The voltage-sensitive potassium channels are controlled by a single type of gate, called the n gate, which is closed at the resting E_m and opens slowly upon depolarization. In response to depolarization, p_{Na} increases dramatically as m gates open, and E_m is driven up near E_{Na}. With a delay, h gates close, restoring p_{Na} to a low level, and n gates open, increasing p_K. As a result, p_{Na}/p_K falls below its normal resting value, and E_m is driven back to near E_K. The resulting repolarization restores the membrane to its resting state.

The behavior of the voltage-dependent sodium and potassium channels can explain (1) why depolarization is the stimulus for generation of an action potential; (2) why action potentials are all-or-none events; (3) how action potentials propagate along nerve fibers; (4) why the membrane potential becomes positive at the peak of the action potential; (5) why the membrane potential is transiently more negative than usual at the end of an action potential; and (6) the existence of a refractory period after a neuron fires an action potential.

6

Voltage-Clamp Experiments

In Chapter 5, we discussed the basic membrane mechanisms underlying the generation of the action potential in a neuron. We saw that all the properties of the action potential could be explained by the actions of voltage-sensitive sodium and potassium channels in the plasma membrane, both of which behave as though there are voltage-activated gates that control permeation of ions through the channel. In this chapter, we will discuss the experimental evidence that gave rise to this scheme for explaining the action potential. The fundamental experiments were performed by A. L. Hodgkin and A. F. Huxley in the period from 1949 to 1952, with the participation of B. Katz in some of the early work. The Hodgkin–Huxley model of the nerve action potential is based on electrical measurements of the flow of ions across the membrane of an axon, using a technique known as voltage clamp. We will start by describing how the voltage clamp works, and then we will discuss the observations Hodgkin and Huxley made and how they arrived at the gated ion channel model discussed in the last chapter.

The Voltage Clamp

We saw in Chapter 5 that the permeability of an excitable cell membrane to sodium and potassium depends on the voltage across the membrane. We also saw that the voltage-induced permeability changes occur at different speeds for the different ionic "gates" on

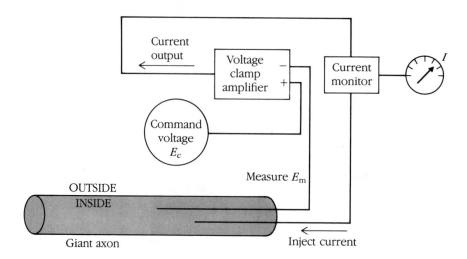

Figure 6-1
Schematic diagram of a voltage-clamp apparatus.

the voltage-sensitive channels. This means that the membrane permeability to sodium, for example, is a function of two variables: voltage and time. Thus, in order to study the permeability in a quantitative way, it is necessary to gain experimental control of one of these two variables. We can then hold that one constant and see how permeability varies as a function of the other variable. The voltage clamp is a recording technique that allows us to accomplish this goal. It holds membrane voltage at a constant value; that is, the membrane potential is "clamped" at a particular level. We can then measure the membrane current flowing at that constant membrane voltage and use the time-course of changes in membrane current as an index of the time-course of the underlying changes in membrane ionic conductance.

A diagram of the apparatus used to voltage clamp an axon is shown in Figure 6-1. Two long, thin wires are threaded longitudinally down the interior of an isolated segment of axon. One wire is used to measure the membrane potential, just as we have done in a number of previous examples using intracellular micro-electrodes; this wire is connected to one of the inputs of the voltage-clamp amplifier. The other wire is used to pass current into the axon and is connected to the output of the voltage-clamp amplifier. The other input of the amplifier is connected to an external voltage source, the command voltage, that is under the experimenter's control. The command voltage is so named because its value determines the value of resting membrane potential that will be maintained by the voltage-clamp amplifier.

The amplifier in the voltage-clamp circuit is wired in such a way that it feeds a current into the axon that is proportional to the difference between the command voltage and the measured membrane potential, $E_C - E_m$. If that difference is zero (that is, if $E_m = E_C$), the amplifier puts out no current, and E_m will remain stable. If E_m does not equal E_C, the amplifier will pass a current into the axon to make the membrane potential move toward the command voltage. For example, if E_m is -70 mV and E_C is -60 mV, then $E_C - E_m$ is a positive number. Because the amplifier passes a current that is proportional to that difference, the current will also be positive. That is, the injected current will move positive charges into the axon and depolarize the membrane toward E_C. This would continue until the membrane potential equals the command potential of -60 mV. On the other hand, if E_C were more negative than E_m, $E_C - E_m$ would be a negative number, and the injected current would be negative. In this case, the current would hyperpolarize the axon until the membrane potential equaled the command voltage.

Measuring Changes in Membrane Ionic Conductance Using the Voltage Clamp

By inserting a current monitor into the output line of the amplifier, we can measure the amount of current that the amplifier is passing to keep the membrane voltage equal to the command voltage. How does this measured current give information about changes in ionic current and, therefore, changes in ionic conductance of the membrane? Consider the effect on the injected current of a sudden increase in the sodium permeability of the membrane, as shown in Figure 6-2A. Suppose we set E_C to be equal to the normal steady-state membrane potential of the cell and turn on the voltage-clamp apparatus. In this situation, E_m is already equal to E_C and the injected current will be zero. Suppose that at some time after we turn on the apparatus, there is a sudden increase in the sodium permeability of the membrane. We know from the discussion of the Goldman equation in Chapter 4 that this would normally cause the membrane potential to take up a new steady-state value closer to the sodium equilibrium potential; that is, the cell would depolarize because of the increase in inward sodium current across the membrane. As shown in Figure 6-2A, however, the voltage-clamp circuit will detect the depolarization as soon as it begins and inject nega-

Figure 6-2
Voltage-clamp currents injected to keep E_m constant in the face of an increase in (A) p_{Na} or (B) p_K of an axon membrane. From top to bottom in both (A) and (B), the traces show the time-course of p_{Na}, p_K, ionic current (sodium in A, potassium in B), and the injected voltage-clamp current. The voltage (above) is held constant between E_{Na} and E_K.

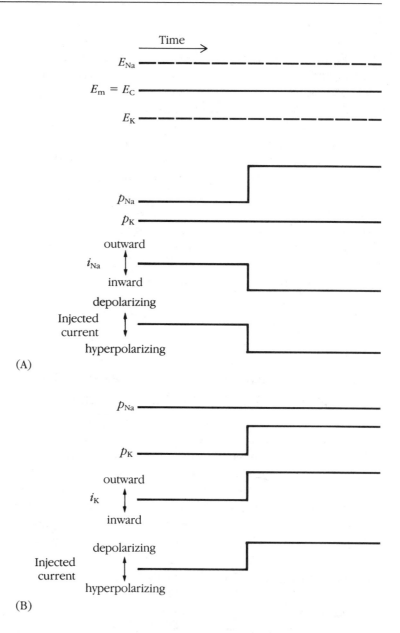

(A)

(B)

tive current into the axon to counter the increased sodium current. The voltage clamp will continue to inject this holding current to maintain E_m at its usual resting value for as long as the increased sodium permeability persists. Thus, the injected current will be equal in magnitude to the increase in sodium current resulting from the increase in sodium permeability. If the potassium permeability, rather than the sodium permeability, increases over its normal resting value, then the voltage-clamp apparatus will respond as shown in Figure 6-2B. In this case, the increased potassium permeability would normally drive E_m more negative, toward E_K, and the cell would hyperpolarize. However, the voltage-clamp amplifier will inject a depolarizing current of the right magnitude to counteract the hyperpolarizing potassium current leaving the cell. The point is that the current injected by the voltage clamp gives a direct measure of the change in ionic current resulting from a change in membrane permeability to an ion.

How do we relate the measured change in membrane current to the underlying change in membrane permeability? Recall from Chapter 4 that the ionic current carried by a particular ion is given by the product of the membrane conductance to that ion and the voltage driving force for that ion, which is the difference between the actual value of membrane potential and the equilibrium potential for the ion. For example, for sodium ions

$$i_{Na} = g_{Na}(E_m - E_{Na}) \qquad (6\text{-}1)$$

Thus, we can calculate g_{Na} from the measured i_{Na} according to the relation

$$g_{Na} = i_{Na}/(E_m - E_{Na}) \qquad (6\text{-}2)$$

In this calculation, E_m is equal to the value set by the voltage clamp, and E_{Na} can be computed from the Nernst equation or measured experimentally by setting E_C to different values and determining the setting that produces no change in ionic current upon a change in g_{Na} (that is, $E_m - E_{Na} = 0$). In this way, it is straightforward to obtain a measure of the time-course of a change in membrane ionic conductance from the time-course of the change in ionic current. As discussed in Chapter 4, conductance is *not* the same as permeability. However, for rapid changes in permeability like those underlying the action potential, we can treat the two as having the same time-course.

The Squid Giant Axon

The experimental arrangement diagrammed in Figure 6-1 was technically feasible only because nature provided neurophysiology with an axon large enough to allow experimenters to thread a pair of wires down the inside. The axon used by Hodgkin and Huxley was the giant axon from the nerve cord of the squid. This axon can

Figure 6-3

Diagram of the current injected by a voltage-clamp amplifier into an axon in response to a voltage step from −70 to −20 mV.

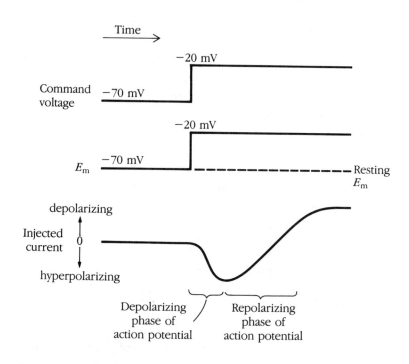

be up to 1 mm in diameter, large enough to be dissected free from the surrounding nerve fibers and subjected to the voltage-clamp procedure described above. The axon is so large that it is possible to squeeze the normal ICF out of the fiber—like toothpaste out of a tube—and replace it with artificial ICF of the experimenter's concoction. This allows the tremendous experimental advantage of being able to control the compositions of both the intracellular and the extracellular fluids.

Ionic Currents Across an Axon Membrane Under Voltage Clamp

The membrane currents flowing in a squid giant axon during a maintained depolarization can be studied in an experiment like that shown in Figure 6-3. In this case, the command voltage to the voltage-clamp amplifier is first set to be equal to the normal resting potential of the axon, which is about −60 to −70 mV. The command voltage is then suddenly stepped to −20 mV, driving the membrane potential rapidly up to the same depolarized value. A depolarization of this magnitude is well above threshold for eliciting an action potential in the axon; however, the voltage-clamp circuit prevents the membrane potential from undergoing the usual sequence of changes that occur during an action potential. The membrane potential remains clamped at -20 mV. What current must the voltage-clamp amplifier inject into the axon in order to keep E_m at -20 mV?

The sodium permeability of the membrane will increase in response to the depolarization and an increased sodium current will enter the axon through the increased membrane conductance to sodium. In the absence of the voltage clamp, this would set up a regenerative depolarization that would drive E_m up near E_{Na}, to about $+50$ mV. In order to counter this further depolarization, the voltage-clamp amplifier must inject a hyperpolarizing current during the strong depolarizing phase of the action potential. With time, however, the sodium permeability of the membrane declines, and the potassium permeability increases in response to the depolarization of the membrane. Normally, this would drive E_m back down near E_K. To counter this tendency and maintain E_m at -20 mV, the voltage clamp then must pass a depolarizing current that is maintained as long as potassium permeability remains elevated. Thus, in response to a depolarizing step above threshold, the membrane of an excitable cell would be expected to show a transient inward current followed by a maintained outward current. The voltage-clamp records of membrane current illustrating this sequence of changes are shown in Figure 6-3.

What was the nature of the evidence that the initial inward current was carried by sodium ions? This was demonstrated by measuring the membrane current resulting from a series of voltage steps of different amplitudes. As we have seen previously, if the clamped value of membrane potential were equal to the sodium equilibrium potential, there would be no driving force for a net sodium current across the membrane. Therefore, if the initial current is carried by sodium ions, that component of the current should disappear when the command voltage is equal to E_{Na}. A sample of membrane current observed in response to a voltage step to E_{Na} is shown in Figure 6-4. The initial component of inward current disappears in this situation, leaving only the late outward current. Hodgkin and Huxley went one step further and systematically varied E_{Na} by altering the external sodium concentration; they found that the membrane potential at which the early current component disappeared was always E_{Na}. This is strong evidence that the inward component of current in response to a depolarization is carried by sodium ions. This notion also agrees with early observations that the membrane potential reached by the

Figure 6-4

Diagram of the current injected by a voltage-clamp amplifier into an axon in response to a voltage step from the normal resting membrane potential to the sodium equilibrium potential. The initial sodium current is absent because there is no driving force for sodium current when E_m equals E_{Na}.

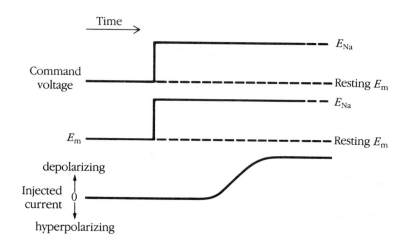

peak of the action potential was strongly influenced by the external sodium concentration.

The two components of membrane current can be separated by comparing the current observed following a voltage step to a particular voltage when that voltage is equal to E_{Na} and when E_{Na} has been moved to another value by altering the external sodium concentration. A specific example is shown in Figure 6-5. In this case, voltage-clamp steps are made to 0 mV in ECF containing normal sodium and in ECF with sodium reduced to be equal with internal sodium concentration. In the normal sodium ECF, E_{Na} will be positive to the command voltage; in the reduced sodium ECF, E_{Na} will equal the command potential and there will be no net sodium current across the membrane. When the observed current in reduced sodium ECF is subtracted from the current in normal ECF, the difference will be the sodium component of membrane current in response to a step depolarization to 0 mV. This isolated sodium current is shown in Figure 6-5C. The membrane currents of Figure 6-5C can be converted to membrane conductance according to equation (6-2), and the result gives the time-course of the membrane sodium and potassium conductances in response to a voltage-clamp step to 0 mV. This procedure can be repeated for a series of different values of command potential and E_{Na}, generating a full characterization of the sodium and potassium conductance changes as a function of both time and membrane voltage. The increase in sodium conductance in response to depolarization is transient, even if the depolarization is maintained. The increasing phase is called **sodium activation**, and the delayed fall is called

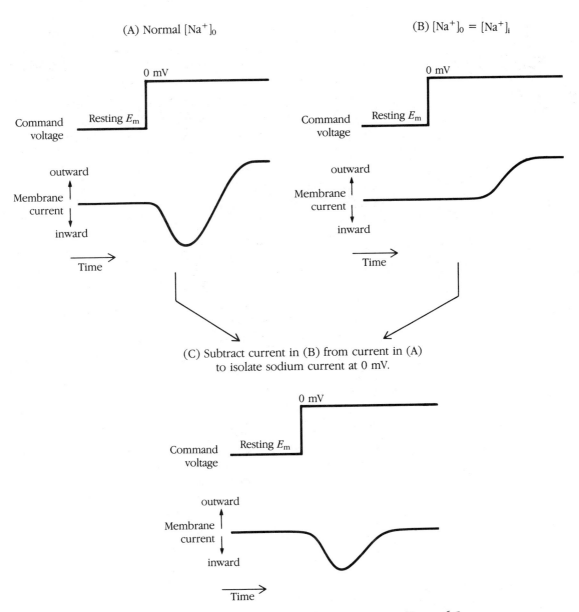

(A) Normal $[Na^+]_o$

(B) $[Na^+]_o = [Na^+]_i$

(C) Subtract current in (B) from current in (A)
to isolate sodium current at 0 mV.

sodium inactivation. We will discuss activation first and return later to the mechanism of inactivation. The onset of the increase in potassium conductance is slower than sodium activation and does not inactivate with maintained depolarization. Thus, at least on the brief time-scale relevant to the action potential, potassium conductance remains high for the duration of the depolarizing voltage step.

Figure 6-5
Procedure for isolating the sodium component of membrane current by varying external sodium concentration to alter the sodium equilibrium potential.

This rather involved procedure has been simplified considerably by the discovery of specific drugs that block the voltage-sensitive sodium channels and other drugs that block the voltage-sensitive potassium channels. The sodium channel blockers most commonly used are the biological toxins tetrodotoxin and saxitoxin. Both seem to interact with specific sites within the aqueous pore of the channel and physically plug the channel to prevent sodium movement. Potassium channel blockers include tetraethylammonium (TEA) and 4-aminopyridine (4-AP). Thus, the isolated behavior of the sodium current could be studied by treating an axon with TEA, while the isolated potassium current could be studied in the presence of tetrodotoxin.

The Gated Ion Channel Model

Membrane Potential and Peak Ionic Conductance

Hodgkin and Huxley discovered that the peak magnitude of the conductance change produced by a depolarizing voltage-clamp step depended on the size of the step. This established the voltage dependence of the sodium and potassium conductances of the axon membrane. The form of this dependence is shown in Figure 6-6 for both the sodium and potassium conductances. Note the steepness of the curves in both cases. For example, a voltage step to -50 mV barely increases g_{Na}, but a step to -30 mV produces a large increase in g_{Na}. Hodgkin and Huxley suggested a simple model that could account for voltage sensitivity of the sodium and potassium conductances. Their model assumes that many individual ion channels, each with a small ionic conductance, determine the behavior of the whole membrane as measured with the voltage-clamp procedure; and that each channel has two conducting states: an open state in which ions are free to cross through the pore, and a closed state in which the pore is blocked. That is, the channels behave as though access to the pore were controlled by a gate. In this scheme, a change in membrane potential alters the probability that a channel will be in the open, conducting state. With depolarization, the probability that a channel is open increases, so that a larger fraction of the total population of channels is open, and the total membrane conductance to that ion increases.

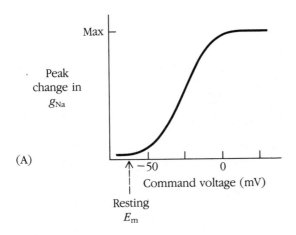

(A)

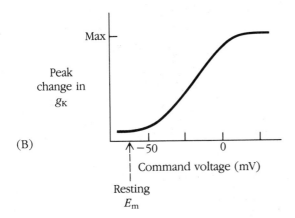

(B)

Figure 6-6
Voltage-dependence of peak
sodium (A) and potassium (B)
conductances as a function of the
amplitude of a maintained voltage
step.

The maximum conductance is reached when all the channels are open, so that further depolarization can have no greater effect.

In order for the conducting state of the channel to depend on transmembrane voltage, some charged entity that is either part of the channel protein or associated with it must control the access of ions to the channel. When the membrane potential is near the resting value, these charged particles are in one state that favors closed channels; when the membrane is depolarized, these charged particles take up a new state that favors opening of the channel. One scheme like this is shown in Figure 6-7. The charged particles are assumed to have a positive charge in Figure 6-7; thus, in the presence a large, inside-negative electric field across the membrane, most of the particles would likely be near the inner face of the membrane. Upon depolarization, however, the dis-

Figure 6-7

A schematic representation of the voltage-sensitive gating of a membrane ion channel. The conducting state of the channel is assumed in this model to depend on the binding of a charged particle to a site on the outer face of the membrane.

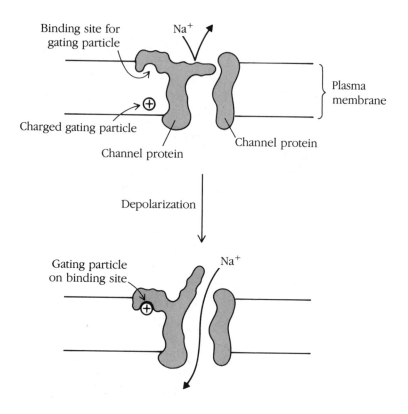

tribution of charged particles within the membrane would become more even, and the fraction of particles on the outside would increase. The channel protein in Figure 6-7 is assumed to have a binding site on the outer edge of the membrane that controls the conformation of the "gating" portion of the channel. When the binding site is unoccupied, the channel is closed; when the site binds one of the positively charged particles (called **gating particles**), the channel opens. Thus, upon depolarization, the fraction of channels with a gating particle on the binding site will increase, as will the total ionic conductance of the membrane. It is important to emphasize that the drawings in Figure 6-7 are illustrative only; it is not clear, for example, that the gating particles are positively charged. Negatively charged particles moving in the opposite direction or a dipole rotating in the membrane could accomplish the same voltage-dependent gating function. The molecular mechanism underlying the change in conducting state of the channel protein is unknown at present. It seems likely, however, that a conformation change related to charge distribution within the membrane is involved.

Kinetics of the Change in Ionic Conductance Following a Step Depolarization

We saw in Chapter 5 that differences in the speed with which the three types of voltage-sensitive gates respond to voltage changes are important in determining the form of the action potential. For instance, the opening of the potassium channels must be delayed with respect to the opening of the sodium channels to avoid wasteful competition between sodium influx and potassium efflux during the depolarizing phase of the action potential. We will now consider how the time-course, or kinetics, of the conductance changes fit into the charged gating particle scheme just presented.

Hodgkin and Huxley assumed that the rate of change in the membrane conductance to an ion following a step depolarization was governed by the rate of redistribution of the gating particles within the membrane. That is, they assumed that the interaction between gating particle and binding site introduced negligible delay into the temporal behavior of the channel. As an example, we will consider the kinetics of opening of the sodium channel following a step depolarization. In formal terms, the movement of gating particles within the membrane can be described by the following first-order kinetic model:

$$m \underset{b_m}{\overset{a_m}{\rightleftharpoons}} (1 - m) \qquad (6\text{-}3)$$

Here, m is the proportion of particles on the outside of the membrane, where they can interact with the binding sites, and $1 - m$ is the proportion of particles on the inside of the membrane. The rate constant, a_m, represents the rate at which particles move from the inner to the outer face of the membrane, and b_m is the rate of reverse movement. Because of the charge on the particles, a step change in the membrane voltage will cause an instantaneous change in the rate constants a_m and b_m. For instance, a step depolarization would increase a_m and decrease b_m, leading to a net increase in m and therefore a decrease in $1 - m$.

The equation governing the rate at which the charges redistribute following a change in membrane potential will be

$$dm/dt = a_m(1 - m) - b_m m \qquad (6\text{-}4)$$

In equation (6-4), dm/dt is the net rate of change of the proportion of particles on the outside face of the membrane. In words,

$a_m(1 - m)$ is the rate at which particles are leaving the inside of the membrane, and $b_m m$ is the rate at which particles are leaving the outside surface; the difference between those two rates is the net rate of change in m. If the distribution of particles is stable—as it would be if E_m had been constant for a long time—the rate at which particles move from inside to outside would equal the rate of movement in the opposite direction, and dm/dt would be zero. If the system is suddenly perturbed by a depolarization, a_m and b_m would change and the balance on the right side of equation (6-4) would be destroyed. If the depolarization is maintained, the rate at which the system will approach a new steady distribution of particles will be governed by equation (6-4).

The solution of a first-order kinetic expression like equation (6-4) is an exponential function; that is, following a step change in membrane voltage, m will approach a new steady value exponentially. The exponential solution can be written

$$m(t) = m_\infty - (m_\infty - m_0)e^{-(a_m+b_m)t} \tag{6-5}$$

This equation states that following a change in membrane potential, m will change exponentially from its initial value (m_0) to its final value (m_∞) at a rate governed by the rate constants (a_m and b_m) for movement of the gating particles at that new value of membrane potential. The behavior of m with time after a depolarization, as expected from equation (6-5), is summarized in Figure 6-8. The number of binding sites occupied by gating particles would be expected to be proportional to m, the fraction of available particles on the outer face of the membrane. Thus, if the occupation of a single binding site causes the channel to open and if the coupling between binding of the gating particle and opening of the channel involves no significant delays, the number of open channels would be expected to follow the same exponential time-course as m after a step depolarization.

Because the total membrane sodium conductance is determined by the number of open sodium channels, sodium conductance measured with a voltage clamp would be expected to be exponential as well, given the assumption of a single gating particle leading to opening of the channel. This prediction, along with the actually observed kinetic behavior of g_{Na}, is diagrammed in Figure 6-9. Unlike the predicted exponential behavior, the rise in g_{Na} actually exhibited a pronounced delay following the voltage step.

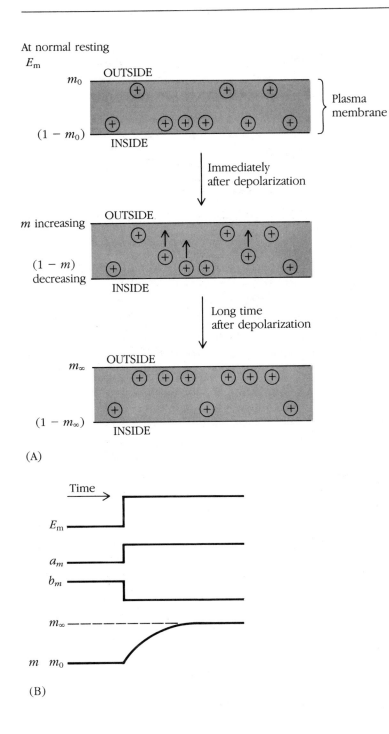

At normal resting E_m

m_0

OUTSIDE

$(1 - m_0)$

INSIDE

Plasma membrane

Immediately after depolarization

m increasing

OUTSIDE

$(1 - m)$ decreasing

INSIDE

Long time after depolarization

m_∞

OUTSIDE

$(1 - m_\infty)$

INSIDE

(A)

Time

E_m

a_m

b_m

m_∞

m m_0

(B)

Figure 6-8

Change in the distribution of sodium channel gating particles after a depolarization of the membrane.

Figure 6-9
Predicted time-course of the change in sodium conductance following a depolarizing step (dashed line), assuming that the proportion of open channels—and hence the total sodium conductance—is directly related to the fraction of gating particles on the outer face of the membrane. The solid line shows the observed change in sodium conductance following a step depolarization.

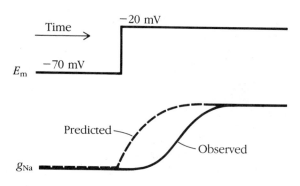

The S-shaped increase in g_{Na} would be explained if more than one binding site must be occupied by gating particles before the channel will open. If the binding to each of several sites is independent, the probability that any one site is occupied will be proportional to m and will thus rise exponentially with time after a step voltage change, as discussed above. The probability that all of a number of sites will be occupied will be the product of the probabilities that each single site will bind a gating particle. That is, if there are two binding sites, the probability that both are occupied will be the product of the probability that site 1 binds a particle and the probability that site 2 binds a particle. Because each of these probabilities is proportional to m, the joint probability that both sites are occupied is proportional to m^2. Similarly, if there were x sites, the probability of channel opening would be proportional to m^x. The actual rise in sodium conductance following a depolarizing step suggested that $x = 3$ for the sodium channel: three binding sites must be occupied by gating particles before the channel will conduct. Thus, the turn-on of g_{Na} following a voltage-clamp step to a particular level of depolarization was proportional to m^3, and the temporal behavior of m was given by equation (6-5).

A similar analysis was carried out for the change in potassium conductance following a step depolarization. The results suggested that $x = 4$ for the voltage-sensitive potassium channel of squid axon membrane. Thus, the gating charges for the potassium channel redistributed after a change in membrane potential according to a relation equivalent to equation (6-4):

$$dn/dt = a_n(1 - n) - b_n n \qquad (6\text{-}6)$$

By analogy with the sodium system, n is the proportion of potassium gating particles on the outside of the membrane, $1 - n$ is the proportion on the inner face of the membrane, and a_n and b_n are the rate constants for particle transition from one face to the other. Equation (6-6) has a solution equivalent to equation (6-5):

$$n(t) = n_\infty - (n_\infty - n_0)e^{-(a_n+b_n)t} \qquad (6\text{-}7)$$

Here, n_0 and n_∞ are the initial and final values of n. The rise in potassium conductance following a step depolarization was found to be proportional to n^4; therefore, the potassium channel behaves as though four binding sites must be occupied by gating particles in order for the gate to open. A major difference between the potassium and the sodium channels is that the rate constants, a_n and b_n, are smaller for potassium channels. That is, the sodium channel gating particles appear to be more mobile than their potassium channel counterparts; this accounts for the greater speed of the sodium channel in opening after a depolarization, which we have seen is a crucial part of the action potential mechanism.

Sodium Inactivation

Recall that the change in sodium conductance following a maintained depolarizing step is transient. We have so far considered only the first part of that change: the increase in sodium conductance called sodium activation. We will now turn to the delayed decline in sodium conductance following depolarization. This delayed decline in conductance is called sodium inactivation. Following along in the vein used in the analysis of sodium and potassium channel opening, Hodgkin and Huxley assumed that sodium inactivation was caused by a voltage-sensitive gating mechanism. They supposed that the conducting state of the sodium channel was controlled by two gates: the activation gate whose opening we discussed above, and the inactivation gate. A diagram of this arrangement is shown in Figure 6-10. Like the activation gate, the inactivation gate is controlled by a charged gating particle; when the binding site on the gate is occupied, the inactivation gate is open. Unlike the activation gate, however, the inactivation gate is normally open, and closes upon depolarization. If we keep the convention of the gating particle being positively charged, this behavior can be modeled by an arrangement with the inactivation gate and its binding site on the inner face of the membrane. Upon depolarization, the probability that a gating particle is on the inner

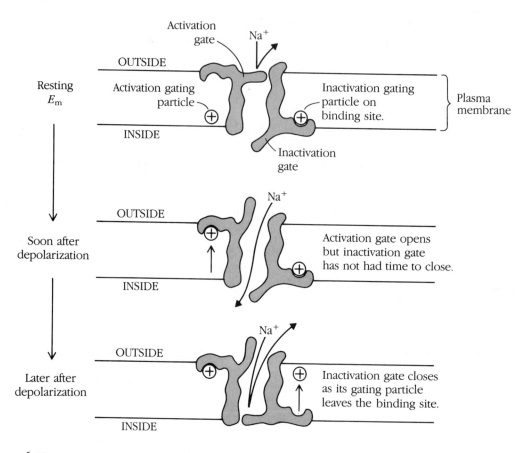

Figure 6-10

Diagram of the sodium channel protein showing effects of both the activation and the inactivation gates.

face decreases, and so the probability that the gate closes will increase.

To study the voltage-dependence of the sodium inactivation process, Hodgkin and Huxley performed the type of experiment illustrated in Figure 6-11. They used a fixed depolarizing test step of a particular amplitude and measured the peak amplitude of the increase in sodium conductance that resulted from the test step. The test depolarization was preceded by a long-duration prepulse whose amplitude could be varied. As shown in Figure 6-11, they found that a depolarizing prepulse reduced the amplitude of the response to the test depolarization, while a hyperpolarizing prepulse increased the size of the test response. This implied that the depolarizing prepulses closed the inactivation gates of some portion of the sodium channels, so that those channels did not conduct even when the activation gates were opened by the subsequent depolarization; therefore, there was a smaller increase in sodium conductance during the test step. The finding that hyperpolarizing

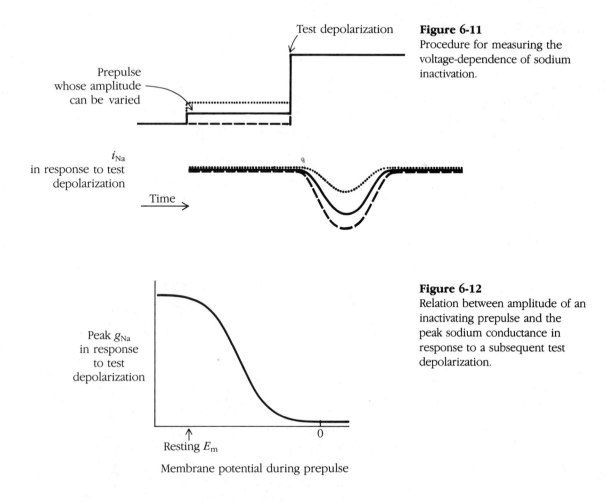

Figure 6-11
Procedure for measuring the voltage-dependence of sodium inactivation.

Figure 6-12
Relation between amplitude of an inactivating prepulse and the peak sodium conductance in response to a subsequent test depolarization.

prepulses increased the test response suggests that some of the inactivation gates of some portion of the sodium channels are already closed at the normal resting potential; increasing E_m causes those gates to open, and the channels are then able to conduct in response to the test depolarization. By varying the amplitude of the prepulse, Hodgkin and Huxley were able to establish the dependence of the inactivation gate on membrane potential. The relation between E_m during the prepulse and the peak sodium conductance during the test depolarization is shown in Figure 6-12. Note that all the inactivation gates close when the membrane potential reaches about 0 mV, and that even a small depolarization can cause a significant reduction in the peak change in sodium conductance.

The time-course of sodium inactivation was studied by varying the duration of the prepulse, rather than its amplitude. With short

Figure 6-13
(A) Procedure for measuring the time-course of sodium inactivation by varying the duration of depolarizing prepulses. (B) Resulting exponential time-course of the closing of the inactivation gate of the sodium channel.

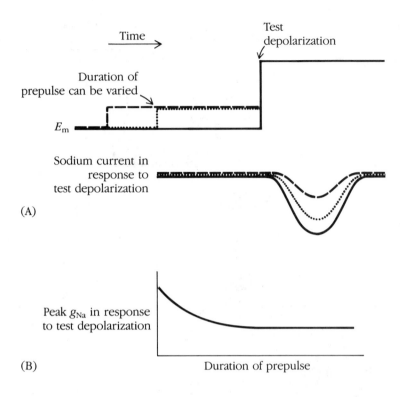

Time

Test depolarization

Duration of prepulse can be varied

E_m

Sodium current in response to test depolarization

(A)

Peak g_{Na} in response to test depolarization

(B)   Duration of prepulse

prepulses, there was not much time for the inactivation gates to close, and the response to the test depolarization was only slightly reduced. With longer prepulses, there was a progressively larger effect. This relation between prepulse duration and peak sodium conductance during the test step is shown in Figure 6-13. It was found that the data were described by a single exponential equation, rather than the powers of exponentials that were necessary to describe the kinetics of sodium and potassium activation. Recall from the discussion of the voltage-dependent opening of the sodium channel that a single exponential is what would be expected if the state of the gate is controlled by a single gating particle. Thus, the closing of the inactivation gate seems to occur when a single particle comes off a single binding site on the gating mechanism. An equation analogous to equations (6-5) and (6-7) can be written to describe the temporal behavior of the inactivation gate:

$$h(t) = h_\infty - (h_\infty - h_0)e^{-(a_n + b_n)t} \qquad (6\text{-}8)$$

In this case, however, the parameter h decreases with depolarization; that is, upon depolarization, h declines exponentially from its original value (h_0) to its final value (h_∞). The rate of that decline is governed by the rate constants, a_h and b_h, for movement of the inactivation gating particle through the membrane. As expected from the discussion in Chapter 5, the closing of the inactivation gate is slower than the operning of the activation gate, implying that the inactivation gating particle is less mobile (rate constants are smaller).

The Temporal Behavior of Sodium and Potassium Conductance

To specify the time-course of the changes in sodium and potassium conductances following a depolarizing voltage-clamp step, it is sufficient to know the behavior of the gating parameters *m, n,* and *h.* In equation form, the sodium and potassium conductances would be given by

$$g_{Na} = \bar{g}_{Na}m^3h \tag{6-9}$$

$$g_K = \bar{g}_K n^4 \tag{6-10}$$

where $\bar{g}_{Na}$ and $\bar{g}_K$ are the maximal sodium and potassium conductances, and *m, n,* and *h* are given by equations (6-5), (6-7), and (6-8), respectively. Thus, following a depolarization, the sodium conductance rises in proportion to the third power of the activation parameter *m* and falls in direct proportion to the decline in the inactivation parameter, *h.* The potassium conductance rises as the fourth power of its activation parameter and does not inactivate. The names used in Chapter 5 for the various voltage-sensitive gates of the potassium and sodium channels derive from the variables chosen by Hodgkin and Huxley to represent these activation and inactivation parameters. The sodium activation gate is called the *m* gate, the sodium inactivation gate the *h* gate, and the potassium gate the *n* gate to reflect the roles of those parameters in equations (6-9) and (6-10).

The surest test of a theory like the Hodgkin and Huxley theory of the action potential is to see if it can quantitatively describe the event it is supposed to explain. Hodgkin and Huxley tested their theory in this way by determining if they could quantitatively reconstruct the action potential of a squid giant axon using the system of equations they derived from their analysis of voltage-clamp data. Because the action potential does not occur under voltage-clamp conditions, this required knowing both the voltage-dependence and the time-dependence of a large number of parameters. This included knowing how the rate constants for all three gating particles and how the maximum values of *h, m,* and *n* depend on the membrane voltage. All of these parameters could be determined experimentally from a complete set of voltage-clamp experiments, allowing Hodgkin and Huxley to calculate the action potential that would occur if their axon were not voltage-clamped. They then compared their calculated action potential with the action potential recorded from the same axon when the voltage-clamp apparatus was switched off. They found that the calculated action potential reproduced all the features of the real one in exquisite detail, confirming that they had covered all the relevant features of the nerve membrane involved in the generation of the action potential.

Gating Currents

Hodgkin and Huxley realized that their scheme for the gating of the sodium and potassium channel predicted that there should be an electrical current flow within the membrane associated with the movement of the charged gating particles. When a step change in membrane potential is made, the charged gating particles redistribute within the membrane; because the movement of charge through space is an electrical current (by definition), this redistribution of charges from one face of the membrane to the other should be measurable as a rapid component of membrane current in response to the voltage clamp. A current of this type flowing within a material is called a **displacement current**. The equipment available to Hodgkin and Huxley was inadequate to detect this small current, however. Almost 20 years later, Armstrong and Bezanilla managed to measure the displacement current associated with the movement of the gating particles.

The procedure for measuring the displacement currents, which have come to be called **gating currents** because of their presumed function in the membrane, is illustrated in Figure 6-14. The basic idea is to start by holding the membrane potential at a hyperpolarized level; this insures that all the gating particles are on the inner face of the membrane (assuming, once again, that the gating particles are positively charged). In addition, all the sodium and potassium currents through the channels are blocked by drugs, like tetrodotoxin and tetraethylammonium. A step is then made to a more hyperpolarized level, say 30 mV more negative. Because all the gating charges are already on the inner face of the membrane,

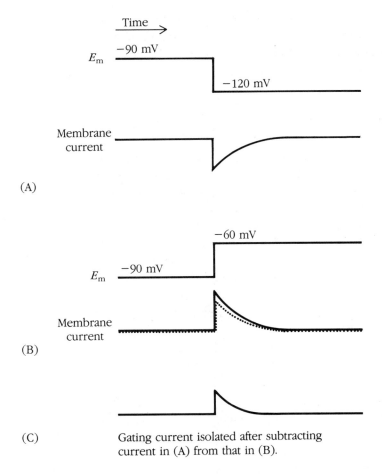

(A)

(B)

(C)   Gating current isolated after subtracting
current in (A) from that in (B).

Figure 6-14
Procedure for isolating the gating
current associated with opening
of voltage-sensitive sodium
channels of an axon membrane.
(A) Membrane voltage is stepped
negative from a hyperpolarized
level. With all ion channels
blocked, the only current flowing
is that required to move the
membrane voltage more negative.
(B) Membrane voltage is stepped
positive from a hyperpolarized
level. The current necessary to
move the potential in the positive
direction (dotted trace) will be
the same amplitude, but opposite
sign, as in (A). In addition, there
will be an extra component of
current in (B) caused by the
movement of the charged gating
particles in response to the
depolarization. This component is
seen in (C) on an expanded
vertical scale.

no displacement current will flow as the result of this hyperpolarizing step. The only current flowing in this situation will be
the rapid influx of negative charge necessary to step the voltage
down. The voltage is then returned to the original hyperpolarized
holding level, and a 30 mV depolarizing step is made. The influx of
positive charge necessary to depolarize by 30 mV will be equal in
magnitude, but opposite in sign, to the influx of negative charge
necessary to make the previous 30 mV hyperpolarizing step. However, the depolarizing step will in addition cause some gating
charges to move from the inner to the outer face of the membrane.
Thus, there will be an extra component of current, due to the
movement of gating charges, in response to the depolarizing step.
By subtracting the current in response to the hyperpolarizing step
from the depolarizing current, this extra gating current can be
isolated. Experiments on this gating current suggest that it has the
right voltage-dependence and other properties to indeed represent

the charge displacement underlying the gating scheme suggested by Hodgkin and Huxley. This is an important piece of evidence validating a basic feature of Hodgkin and Huxley's model of the membrane of excitable cells.

Summary

Hodgkin and Huxley made the fundamental observations on which our current understanding of the ionic basis of the action potential is based. In their experiments, they measured the ionic currents flowing across the membrane of a squid giant axon in response to changes in membrane voltage. This was done using the voltage-clamp apparatus, which provides a means of holding membrane potential constant in the face of changes in the ionic conductance of the axon membrane. By analyzing these ionic currents, Hodgkin and Huxley derived equations specifying both the voltage-dependence and the time-course of changes in sodium and potassium conductance of the membrane. During a maintained depolarization, the sodium conductance increased rapidly, then declined, while potassium conductance showed a delayed but maintained increase. Analysis of the change in sodium conductance suggested that the conducting state of the sodium channel was controlled by a rapidly opening activation gate, called the m gate, and a slowly closing inactivation gate, called the h gate. The gates behave as though they are controlled by charged gating particles that move within the plasma membrane; when the gating particles occupy binding sites associated with the channel gating mechanism, the gates open. The kinetics of the observed gating behavior would be explained by the kinetics of the redistribution of the charged gating particles within the membrane following a step change in the transmembrane potential. The sodium activation gate appears to open when three independent binding sites are occupied by gating particles, while the inactivation gate closes when a single particle leaves a single binding site. The potassium channel is controlled by a single gate, the n gate, which opens when four binding sites are occupied. The rate at which the gating particles redistribute following a depolarization is different for the three types of gate, with sodium activation gating being faster than sodium inactivation or potassium activation. Tiny membrane currents associated with the movement of the charged gating particles within the membrane have been detected.

7

Synaptic Transmission at the Neuromuscular Junction

Chapter 5 was concerned with the ionic basis of the action potential, the electrical signal that carries messages long distances along nerve fibers. Using the patellar reflex as an example, we discussed the mechanism that allows the message that the muscle was stretched to travel along the membrane of the sensory neuron from the sensory endings in the muscle to the termination of the sensory fiber in the spinal cord. This chapter will be concerned with the mechanism by which activity in the motor neuron can be passed along to the cells of the muscle, causing the muscle cells to contract. Chapter 8 will then consider how action potentials in the sensory neuron influence the activity of the motor neuron.

Chemical and Electrical Synapses

The point where activity is transmitted from one nerve cell to another or from a motor neuron to a muscle cell is called a **synapse**. In the patellar reflex, there are two synapses: the one between the sensory neuron and the motor neuron in the spinal cord, and the one between the motor neuron and the cells of the quadriceps muscle. There are two general classes of synapse: electrical synapses and chemical synapses. In both types, there are specialized membrane structures at the point where the input cell, called the **presynaptic cell**, comes into contact with the output cell, called the **postsynaptic cell**.

At a chemical synapse, an action potential in the presynaptic cell causes it to release a chemical substance (called a **neuro-**

transmitter), which diffuses through the extracellular space and changes the membrane potential of the postsynaptic cell. At an electrical synapse, part of a change in membrane potential (such as the depolarization during an action potential) in the presynaptic cell spreads directly to the postsynaptic cell without the action of any intermediary chemical. Both synapses in the patellar reflex, and indeed most synapses in mammalian nervous systems, are chemical synapses.

The Neuromuscular Junction as a Model Chemical Synapse

The best understood chemical synapse is that between a motor neuron and a muscle cell. This synapse is given the special name **neuromuscular junction** (also sometimes called the **myoneural junction**). Although the fine details may differ somewhat, the basic scheme that describes the functioning of the neuromuscular junction applies to all chemical synapses, as far as is known. Therefore, this chapter will concentrate on the characteristics of this special synapse at the end of the patellar reflex. In the next chapter, we will consider some of the differences between the synapse at the neuromuscular junction and synapses in the central nervous system, like that between the sensory and motor neurons in the spinal cord.

Figure 7-1
Sequence of events during transmission at a chemical synapse.

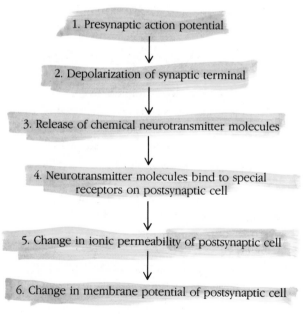

1. Presynaptic action potential

2. Depolarization of synaptic terminal

3. Release of chemical neurotransmitter molecules

4. Neurotransmitter molecules bind to special receptors on postsynaptic cell

5. Change in ionic permeability of postsynaptic cell

6. Change in membrane potential of postsynaptic cell

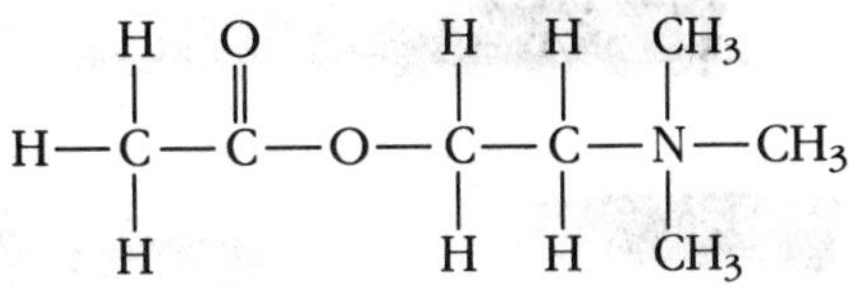

$$H-\underset{\underset{H}{|}}{\overset{\overset{H}{|}}{C}}-\overset{\overset{O}{\|}}{C}-O-\underset{\underset{H}{|}}{\overset{\overset{H}{|}}{C}}-\underset{\underset{H}{|}}{\overset{\overset{H}{|}}{C}}-\underset{\underset{CH_3}{|}}{\overset{\overset{CH_3}{|}}{N}}-CH_3$$

Figure 7-2
Chemical structure of acetylcholine, the chemical neurotransmitter at the neuromuscular junction.

Transmission at a Chemical Synapse

The sequence of events during neuromuscular synaptic transmission is summarized in Figure 7-1. When an action potential arrives at the end of the motor neuron nerve fiber, it invades a specialized structure called the **synaptic terminal**. Depolarization of the synaptic terminal induces release of a chemical messenger, which is stored inside the terminal. At the vertebrate neuromuscular junction, this chemical messenger is **acetylcholine**, whose chemical structure is shown in Figure 7-2. The acetylcholine (abbreviated ACh) diffuses across the space separating the presynaptic motor neuron terminal from the postsynaptic muscle cell and alters the ionic permeability of the muscle cell membrane. The change in ionic permeability produces a depolarization of the muscle cell membrane. The remainder of this chapter will be concerned with a detailed description of this basic sequence of events. The linkage between depolarization of the muscle cell and contraction of the muscle will be discussed later, in Chapter 9.

Presynaptic Action Potential and Acetylcholine Release

The occurrence of an action potential in the synaptic terminal is the trigger for ACh release. It has been demonstrated that the depolarization during the action potential is the necessary stimulus for transmitter release; any experimental manipulation that depolarizes the synaptic terminal causes ACh to be released. This coupling between depolarization and release is not direct, however. The signal that mediates this coupling is the influx into the synaptic terminal of an ion in the ECF that we have ignored to this point—calcium.

Calcium is present at low concentration in ECF ($1 - 2$ mM) and is not important in resting membrane potentials or nerve action potentials; therefore, it was ignored in discussions in previous chapters. However, calcium ions must be present in the ECF in

Figure 7-3

Sequence of events between the arrival of an action potential at a synaptic terminal and the release of chemical transmitter.

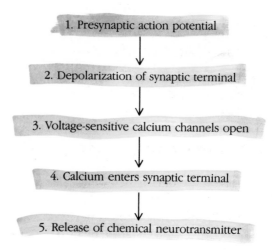

1. Presynaptic action potential

2. Depolarization of synaptic terminal

3. Voltage-sensitive calcium channels open

4. Calcium enters synaptic terminal

5. Release of chemical neurotransmitter

order for release of chemical neurotransmitters to occur. If calcium ions are removed from the ECF, depolarization of the synaptic terminal can no longer induce release of ACh. Thus, depolarization causes external calcium ions to enter the synaptic terminal, and the calcium in turn causes ACh to be released from the terminal.

What mechanism underlies the linkage between depolarization of the terminal and influx of calcium ions? As we've seen in earlier chapters, ions cross membranes through specialized transmembrane channels, and calcium ions are no different in this regard. The membrane of the synaptic terminal contains calcium channels that are closed as long as E_m is near its normal resting level. These channels are similar in behavior to the voltage-dependent potassium channels of nerve membrane; they open upon depolarization and close again when the membrane potential repolarizes. Thus, when an action potential invades the synaptic terminal, the calcium permeability of the membrane increases during the depolarizing phase of the action potential and declines again as membrane potential returns to normal.

Although the external calcium concentration is small (1 − 2 mM), the internal concentration of calcium ions that are free to diffuse across the plasma membrane is much smaller ($< 10^{-6} M$). From the Nernst equation, then, the equilibrium potential for calcium would be expected to be positive. Therefore, both the concentration and electrical gradients for calcium drive calcium into the terminal, and when calcium permeability increases, there will be an influx of calcium. During a presynaptic action potential, there is a spike of calcium entry into the terminal, resulting in release of neurotransmitter into the extracellular space. This sequence is summarized in Figure 7-3.

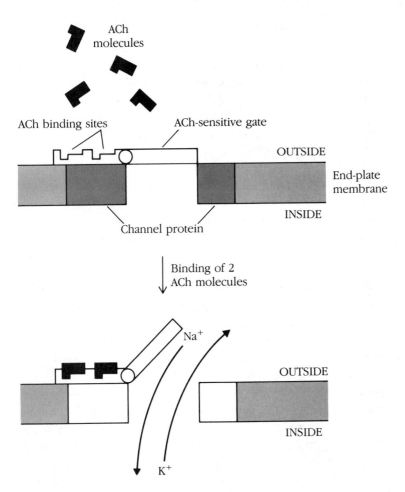

Figure 7-4
Schematic representation of the behavior of the acetylcholine-sensitive channel in the end-plate membrane.

Effect of ACh on the Muscle Cell

The goal of synaptic transmission at the neuromuscular junction is to cause the muscle cell to contract. Acetylcholine released from the synaptic terminal accomplishes this goal by depolarizing the muscle cell. Because muscle cells are excitable cells like neurons, this depolarization will set in motion an all-or-none, propagating action potential if the depolarization exceeds threshold. The coupling between the muscle action potential and contraction will be the subject of Chapter 9. This section will discuss the effect of ACh on the muscle cell membrane.

The region of muscle membrane where synaptic contact is made is called the **end-plate** region, and it possesses special characteristics. In particular, the end-plate membrane is rich in a trans-membrane protein that acts as an ionic channel. Unlike the voltage-dependent channels discussed in Chapter 5, however, this channel is little affected by membrane potential. Instead, this channel is sensitive to ACh: it opens when it binds ACh. Thus, ACh released from the synaptic terminal diffuses across the synaptic cleft to the muscle membrane, where it combines with specific receptors associated with the ionic channel. As shown schematically in Figure 7-4, the gate on the channel is closed in the absence of ACh. When the receptor sites are occupied, however, the protein undergoes a conformational change, the gate opens, and the channel allows ions across the membrane. Current evidence indicates that two ACh molecules must bind to the channel in order for the gate to open. The ACh-binding site is highly specific; only ACh or a small number of structurally related compounds can bind to the site and cause the channel to open.

The ACh-activated channel of the muscle end-plate allows both sodium and potassium to cross the membrane about equally well. Thus, when ACh is present, the membrane permeability to both sodium and potassium increases. How can such a permeability increase produce a depolarization of the muscle cell? To see this, consider the situation diagrammed in Figure 7-5. Recall from Chapter 4 that membrane potential depends on the relative sodium and potassium permeabilities of the membrane (the Goldman equation). For the cell of Figure 7-5, p_{Na}/p_K is 0.02 at rest and E_m would be about -74 mV, assuming typical ECF and ICF (Table 1-1). In the presence of ACh, however, p_{Na} and p_K increase by equal amounts; p_{Na}/p_K increases to 0.51 and E_m depolarizes to about -17 mV.

In a muscle cell, the situation is somewhat more complicated than in Figure 7-5 because the ACh-activated channels are spatially restricted to the end-plate region. Thus, the permeability increase occurs in only part of the cell membrane. In such a situation, the result would be quantitatively different from but qualitatively similar to the example in Figure 7-5.

The ACh-activated channels are packed densely in the muscle end-plate region, and this provides a good opportunity to visualize the protein molecules that form the channel. Because the protein molecules are very small, however, it is only at high power under the electron microscope that the ACh-receptor and its channel can be discerned. A picture of a postsynaptic membrane examined in this way is shown in Figure 7-6. The membrane is studded with ring-shaped particles that are found only at the region of synaptic contact. These particles have been isolated from the postsynaptic membrane and chemically identified as the ACh-binding receptor molecule and its associated channel. The hole in the middle is

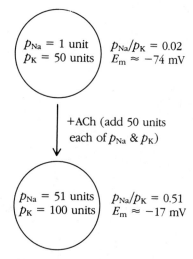

Figure 7-5

Opening a channel that allows both potassium and sodium to cross the membrane results in a higher value for p_{Na}/p_K and causes depolarization.

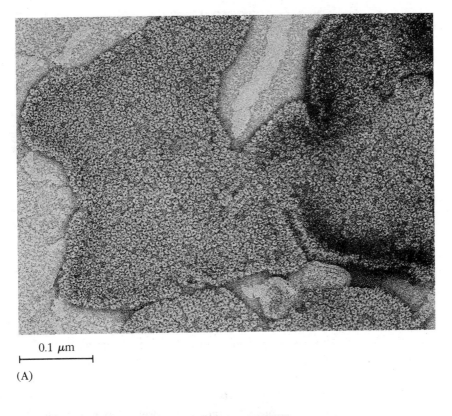

0.1 μm

(A)

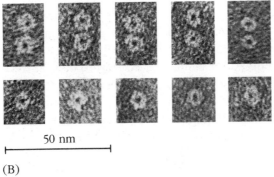

50 nm

(B)

Figure 7-6
(A) A view through the electron microscope at the face of the postsynaptic membrane of the electric organ of the electrical skate, *Torpedo*. This organ, which is a rich source of acetylcholine receptors for biochemical study, is a specialized type of muscle tissue. The membrane particles are the acetylcholine-activated channels of the postsynaptic membrane. (B) Several views of individual acetylcholine receptors that have been chemically isolated from preparations like that in (A), then placed in artificial membranes. [Courtesy of J. Cartaud of the Institut Jacques Monod, Université Paris.]

probably the aqueous pore through which the sodium and potassium ions cross the membrane. The ACh-activated channels have been chemically isolated from muscle membrane, and can be inserted into artificial membranes as in Figure 7-6B. Biochemical analysis of the structure of the channel protein is well underway, and it is likely that the molecular basis of its functioning will soon be understood.

Neurotransmitter Release

We will now return to the synaptic terminal for a more detailed examination of the mechanism of neurotransmitter release. Acetylcholine is released from the motor nerve terminal in quanta consisting of many molecules. Thus, the basic unit of release is not a single molecule of ACh, but the quantum. At the neuromuscular junction, it is estimated that a single quantum of ACh contains about 10,000 molecules. An individual quantum is either released all together or not released at all. The release of ACh during neuromuscular transmission can be thought of as the sudden appearance of a "puff" of ACh molecules in the extracellular space as the entire contents of a quantum is released. A single presynaptic action potential normally causes the release of more than a hundred quanta from the synaptic terminal.

The original suggestion that ACh is released in multimolecular quanta was made on the basis of a statistical analysis of the response of the postsynaptic muscle cell to action potentials in the presynaptic motor neuron. This analysis was first carried out by P. Fatt and B. Katz, and it initiated a series of studies by Katz and co-workers that gave rise to the basic scheme for chemical neurotransmission presented in this chapter. Experimentally, the analysis was accomplished by reducing the extracellular calcium concentration to the point where the influx of calcium ions into the synaptic terminal during an action potential was much less than usual. Under these conditions, a single presynaptic action potential released on average only one or two quanta of ACh instead of more than a hundred. Examples of end-plate potentials recorded in a muscle cell in response to a series of presynaptic action potentials are shown in Figure 7-7. Because only a small number of quanta are released per action potential, the end-plate potentials in the reduced calcium ECF are much smaller than usual and do not reach threshold for generating an action potential in the muscle cell. Notice that the amplitude of the depolarization of the muscle cell fluctuates considerably over the series of presynaptic action potentials: sometimes there was a large response and other times there was no response at all. Fatt and Katz measured a large number of such responses and found that the amplitudes clustered around particular values that were integral multiples of the smallest observed response. For example, as shown in Figure 7-7B, there might be a cluster of responses that were 1 mV in amplitude,

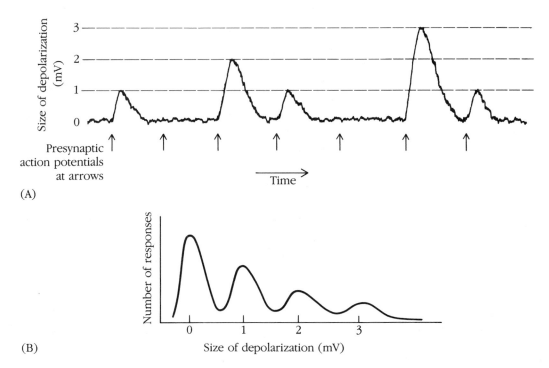

(A)

(B)

Figure 7-7
Quantized responses of muscle cell to action potentials in the presynaptic motor neuron. Arrows give timing of the presynaptic action potentials. The graph in (B) shows the peak response amplitudes that might be recorded in response to a series of several hundred presynaptic action potentials as in (A).

another cluster at 2 mV, and another at 3 mV. This indicates that the response was quantized in irreducible units of 1 mV, and that the presynaptic action potential released ACh in corresponding quantal units. Thus, a given presynaptic action potential might cause release of 3, 2, 1 or no quanta, but not 0.5 or 1.5 quanta.

Fatt and Katz also observed that there were occasional, small depolarizations that occurred in the absence of any presynaptic action potential. They noted that these spontaneous depolarizations always had the same amplitude as the single quantum response produced by a presynaptic action potential in low-calcium ECF. That is, if the irreducible unit of evoked muscle response was 1 mV, then the spontaneous events also were about 1 mV in amplitude. Figure 7-8 shows several of these spontaneous depolarizations recorded inside a muscle cell. These events are called miniature end-plate potentials, and are assumed to result from spontaneous release of single quanta of ACh from the synaptic terminal. Under normal conditions, these spontaneous events occur at a low rate—about 1 or 2 per second; however, any manipulation that depolarizes the nerve terminal increases their rate of occurrence, confirming that their source is the process that couples de-

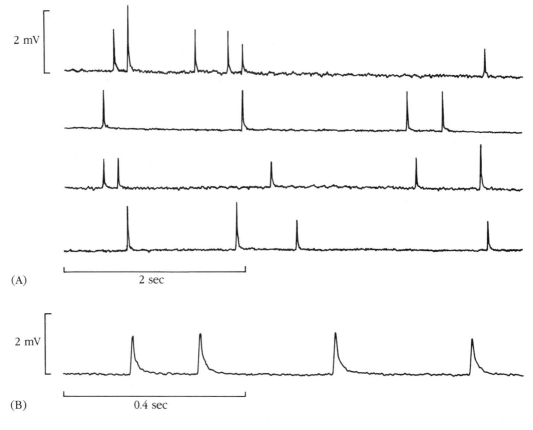

(A)

2 mV

2 sec

(B)

2 mV

0.4 sec

Figure 7-8

Spontaneous miniature end-plate potentials recorded from the end-plate region of a muscle cell. These randomly occurring small depolarizations of the muscle cell are caused by spontaneous release of single quanta of acetylcholine from the synaptic terminal of the motor neuron. (A) Four 5-sec samples of muscle cell E_m, measured via an intracellular microelectrode. The spontaneous depolarizations occur at a rate of approximately one per second. (B) Spontaneous miniature end-plate potentials viewed on an expanded time-scale to show the shape of the events more clearly. [Unpublished data by G. Matthews.]

polarization to quantal ACh release during the normal functioning of the nerve terminal.

The Vesicle Hypothesis of Quantal Transmitter Release

To understand the basis of the packaging of ACh in quanta, it is necessary to look at the structure of the synaptic terminal, which is shown schematically in Figure 7-9. The terminal contains a large number of tiny, membrane-bound structures called **synaptic vesicles**. These vesicles contain ACh, and it is natural to assume that they represent the packets of ACh that are released in response to a presynaptic action potential. Indeed, these vesicles are depleted by any manipulation, such as prolonged depolarization or firing of large numbers of action potentials, which causes release of large amounts of ACh. It is now generally accepted that release of ACh is accomplished by the fusion of the vesicle membrane with the

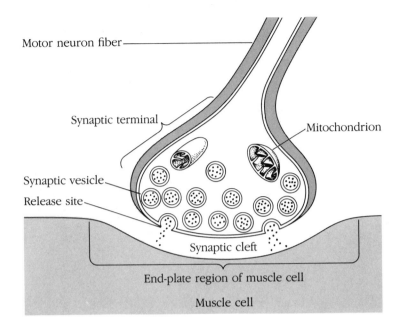

Motor neuron fiber

Synaptic terminal

Mitochondrion

Synaptic vesicle

Release site

Synaptic cleft

End-plate region of muscle cell

Muscle cell

Figure 7-9
The structure of the region of synaptic contact at the neuromuscular junction.

plasma membrane of the terminal, so that the contents of the vesicle are dumped into the extracellular space between the terminal and the muscle cell. The vesicles do not fuse with the plasma membrane just anywhere; rather, they apparently fuse only at specialized membrane regions, called **release sites** or **active zones**, that are found only on the membrane face opposite the postsynaptic muscle cell. Thus, quanta of ACh are released only into the narrow space, the **synaptic cleft**, separating the pre- and postsynaptic cells. A close-up, schematic view of the fusion of a synaptic vesicle with the plasma membrane is shown in Figure 7-10. With freeze-fracture electron microscopy, the active zone of the presynaptic terminal appears as a double row of large membrane particles, which are probably membrane proteins involved in the fusion between the membrane of the synaptic vesicle and the presynaptic plasma membrane. Examples of these active zone particles can be seen in Figure 7-11.

Elegant evidence for vesicle fusion as the mechanism of ACh release was provided by anatomical experiments by Heuser, Reese, Dennis, Jan, Jan, and Evans. In these experiments, a muscle and its attached nerve were placed in an apparatus that could very rapidly freeze the nerve and muscle. Heuser and his coworkers then literally froze the release process at the instant just after arrival of an action potential in the synaptic terminal and examined the frozen

Figure 7-10
Close-up view of the synaptic region during release of acetylcholine at the neuromuscular junction.

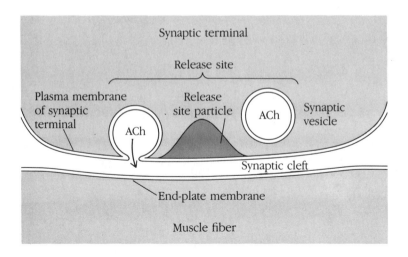

tissue in the electron microscope, using the freeze-fracture technique discussed in Chapter 1. They saw what appeared to be synaptic vesicles in the process of fusing with the plasma membrane, as shown in Figure 7-11. The fusing vesicles appeared as ice-filled pits or depressions in the presynaptic membrane, lined up along the presynaptic release sites. They saw such fusion only when ACh release should have been occurring, not before or after the action potential in the terminal. Further, the fusion occurred only when calcium was present in the ECF, which we have seen is prerequisite for release to occur.

One of the great remaining mysteries in the mechanism of synaptic transmission is the linkage between calcium entry and vesicle fusion. No one yet understands how an increase in internal calcium increases the probability that a vesicle will fuse with the plasma membrane of the synaptic terminal. It is likely that there are calcium-binding molecules associated with the release sites and

Figure 7-11
Electronmicrographs of the freeze-fractured face of a presynaptic terminal at the neuromuscular junction. (A) An unstimulated nerve terminal. Note the double row of particles defining a presynaptic release site or active zone (az). The arrow points to what appears to be a synaptic vesicle spontaneously fusing with the presynaptic membrane. Such spontaneous fusions presumably underlie the spontaneous miniature end-plate potentials shown in Figure 7-8. The arrowhead at the left points to a synaptic vesicle visible in a region where the membrane fractured all the way through to reveal a portion of the intracellular fluid. (B) A higher-power view of an active zone of a nerve terminal frozen during release of acetylcholine stimulated by presynaptic action potentials. The ice-filled depressions arrayed along either side of the active zone correspond to regions where synaptic vesicles are in the process of fusing with the presynaptic membrane. [Reproduced, with permission, from C.-P. Ko, *J. Cell Biol.* 98(1984):1685–1695.]

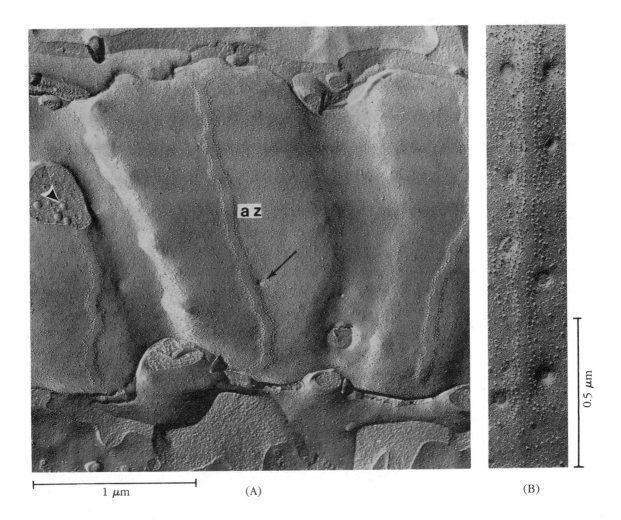

1 μm   (A)   (B)

0.5 μm

with the vesicle membrane and that binding of calcium somehow alters the interactions among these molecules. The rows of membrane particles seen in freeze-fracture views of the presynaptic release sites (Figure 7-11) may correspond to these molecules that regulate the interaction between the vesicles and the presynaptic membrane. However, the molecular mechanism remains unknown.

Recycling of Vesicle Membrane

If the membranes of synaptic vesicles fused with the plasma membrane of the terminal during transmitter release, we might expect the area of the terminal membrane to increase with use. Indeed,

Figure 7-12
The recycling of vesicle membrane in the presynaptic terminal at the neuromuscular junction.

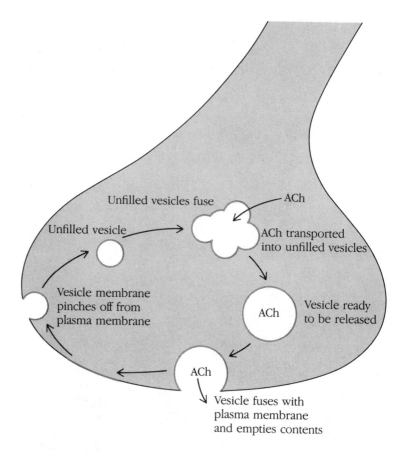

over the life span of an animal, millions of synaptic vesicles might fuse with the terminal membrane, so that the terminal might become huge. However, this does not happen because the vesicle membrane does not remain part of the plasma membrane; instead, Heuser and Reese found that the fused vesicles are recycled. The scheme is summarized in Figure 7-12. After fusion, the vesicles pinch off again from the plasma membrane, are refilled with ACh, and are ready to be used again to transfer neurotransmitter into the synaptic cleft.

Inactivation of Released Acetylcholine

We have seen how ACh is released from the synaptic terminal and how it depolarizes the postsynaptic muscle cell. How is the action of ACh terminated so that the end-plate region returns to its resting state? The answer is that there is another specialized ACh-binding protein in the end-plate region. This protein is the enzyme acetylcholinesterase, which splits ACh into acetate and choline. Because neither acetate nor choline can bind to and activate the ACh-activated channel, the acetylcholinesterase effectively halts the action of any ACh it encounters.

When a puff of ACh is released in response to an action potential in the synaptic terminal, the concentration of ACh in the synaptic cleft abruptly rises. Some of the released ACh molecules will bind to ACh-activated channels, causing them to open and increasing the sodium and potassium permeability of the end-plate membrane; other ACh molecules will bind to acetylcholinesterase and be inactivated. Even though the binding of ACh to the postsynaptic channel is highly specific, it is readily reversible; the binding typically lasts for only about 1 msec. When an ACh molecule comes off a gate, the channel closes. The newly freed ACh molecule may then bind again to an ACh-activated channel, or it might bind to acetylcholinesterase and be inactivated. With time following release of the puff, the concentration of ACh in the cleft will fall until all of the released ACh has been split into acetate and choline.

It would be wasteful if the choline resulting from inactivation of ACh were lost and had to be replaced with fresh choline from inside the presynaptic cell. This potential waste is avoided because most of the choline is taken back up into the synaptic terminal, where it is reassembled into ACh by the enzyme choline acetyltransferase. Thus, both the vesicle membrane (the packaging material of the quantum) and the released neurotransmitter (the contents of the quantum) are effectively recycled by the presynaptic terminal.

Recording the Electrical Current Flowing Through a Single Acetylcholine-Activated Ion Channel

Throughout our discussion of the membrane properties of excitable cells, we have made extensive use of the notion of ions crossing the membrane through specific pores or channels. For example, we saw that the effect of acetylcholine on the muscle membrane is mediated via ion channels in the postsynaptic membrane that open in the presence of acetylcholine. As discussed in Chapters

Figure 7-13

Schematic illustration of the procedure for recording the current through a single acetylcholine-activated channel in a cell membrane. A micropipette with a tip diameter of 1 to 2 μm is placed on the external surface of the membrane. A tight electrical seal is made between the membrane and the glass of the micropipette, so that a resistance greater than 10^{10} Ω is imposed in the extracellular path for current flow through the channel. When a channel in the patch of membrane inside the micropipette opens, a current-sensing amplifier connected to the interior of the pipette detects the minute current flow.

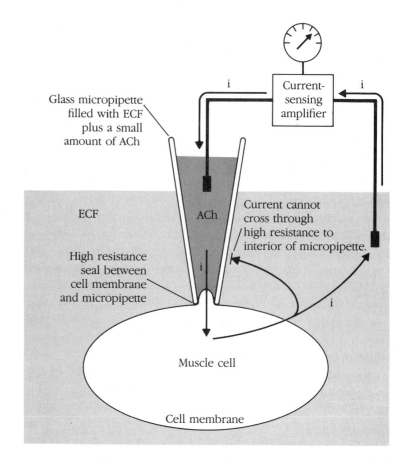

4 and 6, the flow of ions across the cell membrane constitutes a transmembrane electrical current that can be measured with electrical techniques like the voltage clamp. Recently, a new technique was developed by Neher and Sakmann to record transmembrane ionic currents, and the technique has sufficient resolution to measure the minute electrical current flowing through a single open ion channel. The technique is called the **patch clamp**, and it is illustrated in Figure 7-13.

The basic idea behind the patch clamp is to isolate electrically a small patch of cell membrane that contains only a few ionic channels. The electrical isolation is achieved by placing a specially constructed miniature glass pipette in close contact with the membrane. When one of the ion channels in the isolated patch opens, electrical current flows across the membrane; in the case of the acetylcholine-activated channel that current would be a net inward (that is, depolarizing) current under normal conditions. We know from the basic properties of electricity that current must flow in a complete circuit. As shown in Figure 7-13, the return current path through the extracellular space is broken by the presence of the glass pipette; there is a high electrical resistance imposed by the seal between the cell membrane and the pipette. Under these conditions, the ionic current through the open channel is forced to complete its circuit through the current-sensing amplifier connected to the interior of the pipette. In order for the patch-clamp technique to achieve sufficient sensitivity to measure the current through a single channel, the electrical resistance between the interior of the patch pipette and the extracellular space must be greater than about 10^9 Ω, which is a very large resistance indeed. Fortunately for neurophysiologists, it is possible to achieve resistances greater than 10^{10} Ω.

Using the patch clamp, it is possible to record the current through acetylcholine-activated channels of the postsynaptic membrane of muscle cells by placing a small amount of acetylcholine (or structurally related compounds that are recognized by the receptors on the gate) inside the patch pipette. As shown in Figure 7-4, when the receptors are occupied, the gate opens and the channel allows ions to cross the membrane. Schematically, then, we might expect to record an electrical current like that shown in Figure 7-14A when the channel opens. There would be a rapid step of inward current that occurs as the gate opens, the current would be maintained at a constant level for as long as the channel is open, and the step would terminate when acetylcholine unbinds from one of the receptor sites, causing the gate to close. If a second channel opens while the first is still open, the two currents simply add to produce a current twice as large as the single-channel current. This is also shown in Figure 7-14A.

Actual patch-clamp recordings of currents through single acetylcholine-activated channels of human muscle cells are shown in Figure 7-14B. These records show that the currents through the channels are the rectangular events expected from the simple gating scheme of Figure 7-4. Experiments like that of Figure 7-14B confirm directly the view of ion permeation and channel gating that we have used to explain the electrical behavior of the membranes of excitable cells: the gated ion channels carrying electrical current across the plasma membrane are not just figments of the neurophysiologist's imagination. The development of the patch-clamp technique has led to a flurry of new information about ion channels of all types; for example, the currents flowing through single

Figure 7-14
(A) The current that would be expected to flow through a single acetylcholine-activated channel if the conducting state of the channel is controlled by a single gate that is either open all the way or closed all the way. When ACh binds to the gating portion of the channel, the channel opens and there is a stepwise pulse of inward current flowing through the channel. When ACh unbinds, the channel closes and the current abruptly disappears.
(B) Actual recordings of currents flowing through single acetylcholine-activated channels of human muscle cells maintained in cell culture. Note that at the asterisk two channels opened simultaneously, and their currents added independently. These recordings show the fundamental postsynaptic events that underlie the interaction between the nervous system and the muscles in the human body. [Reproduced, with permission, from M. B. Jackson, H. Lecar, V. Askanas, and W. K. Engel, *J. Neurosci.* 2(1982):1465–1473.]

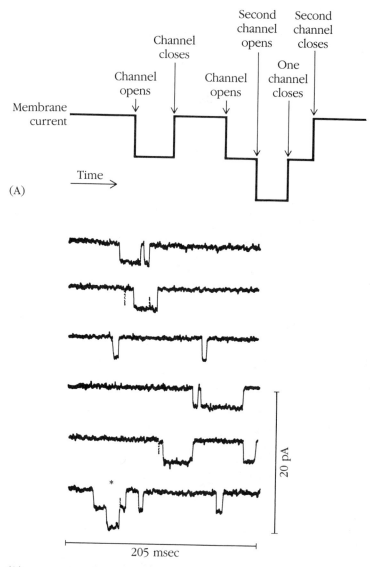

voltage-sensitive sodium and potassium channels that underlie the action potential (see Chapters 5 and 6) have also been observed using this technique.

Summary

The sequence of events during synaptic transmission at the neuromuscular junction is summarized in Figure 7-15. The depolari-

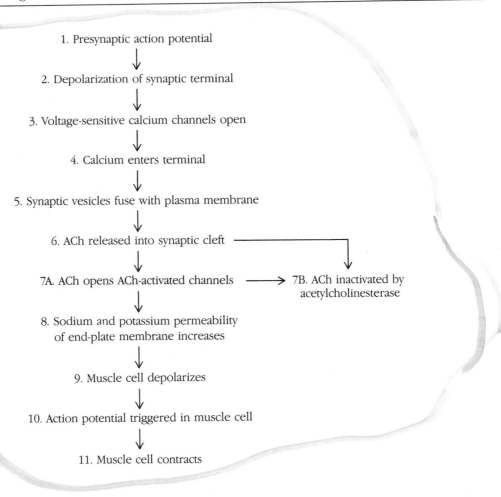

1. Presynaptic action potential

↓

2. Depolarization of synaptic terminal

↓

3. Voltage-sensitive calcium channels open

↓

4. Calcium enters terminal

↓

5. Synaptic vesicles fuse with plasma membrane

↓

6. ACh released into synaptic cleft ⟶

↓

7A. ACh opens ACh-activated channels ⟶ 7B. ACh inactivated by acetylcholinesterase

↓

8. Sodium and potassium permeability of end-plate membrane increases

↓

9. Muscle cell depolarizes

↓

10. Action potential triggered in muscle cell

↓

11. Muscle cell contracts

Figure 7-15
Summary of the sequence of events during synaptic transmission at the neuromuscular junction.

zation produced by an action potential in the synaptic terminal opens voltage-dependent calcium channels in the terminal membrane. Calcium ions enter the terminal down their concentration and electrical gradients, inducing synaptic vesicles filled with acetylcholine to fuse with the plasma membrane facing the muscle cell. The ACh is thereby dumped into the synaptic cleft, and some of it diffuses to the muscle membrane and combines with specific receptors on ACh-activated channels in the muscle membrane. When ACh is bound, the channel opens and allows sodium and potassium ions to cross the membrane. This depolarizes the muscle membrane and triggers an all-or-none action potential in the muscle cell. The action of ACh is terminated by the enzyme acetylcholinesterase, which splits ACh into acetate and choline.

8

Synaptic Transmission in the Central Nervous System

Chemical synapses between neurons operate according to the same general principles as the synapse between a motor neuron and a muscle cell discussed in the previous chapter. Thus, in the patellar reflex, an action potential in the sensory neuron depolarizes the quadriceps motor neuron through a sequence of events similar to that at the neuromuscular junction. There are, however, some major differences between neuron-to-neuron synapses and the neuron-to-muscle synapse discussed previously. This chapter will consider some of those differences, again with specific reference to the patellar reflex.

Excitatory and Inhibitory Synapses

At the neuromuscular junction, acetylcholine depolarizes the muscle cell, causing it to fire an action potential. Synapses of this type are called **excitatory synapses** because the neurotransmitter brings the membrane potential of the postsynaptic cell toward the threshold for firing an action potential and thus tends to "excite" the postsynaptic cell. The synapse between the sensory neuron and the quadriceps motor neuron in the patellar reflex is an example of an excitatory synapse between two neurons. Synapses between neurons are not always excitatory, however. At **inhibitory synapses**, the neurotransmitter acts to keep the membrane potential of the postsynaptic cell more negative than the threshold potential; because this tends to prevent the postsynaptic neuron from firing an action potential, the postsynaptic cell is "inhibited" by the re-

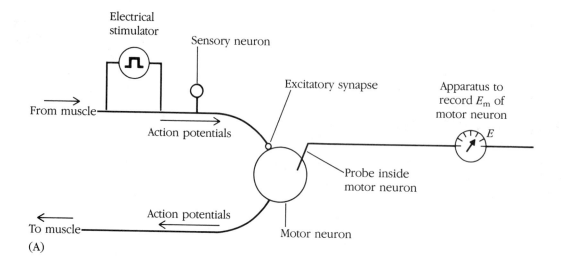

(A)

Figure 8-1

Synaptic transmission at an excitatory synapse between two neurons. (A) Experimental arrangement for examining transmission between a sensory and a motor neuron in the patellar reflex loop.
(B) Responses of the postsynaptic motor neuron to action potentials in the presynaptic sensory neuron. At the upward arrows, action potentials are triggered in the presynaptic neuron by an electrical stimulus.

lease of the inhibitory neurotransmitter. The fact that not all synapses in the nervous system are excitatory is one major difference between synaptic transmission at the neuromuscular junction and synaptic transmission in the nervous system in general.

We will return to a discussion of inhibitory synapses later in this chapter. At this point, the discussion will center on the properties of excitatory synaptic transmission between neurons in the nervous system.

Excitatory Synaptic Transmission Between Neurons

The synapse at the neuromuscular junction is unusual in one important regard: a single action potential in the presynaptic motor neuron produces a sufficiently large depolarization in the postsynaptic muscle cell to trigger a postsynaptic action potential. Such a synapse is called a one-for-one synapse because one action potential appears in the output cell for each action potential in the input cell. Most synapses between neurons are not this strong. Instead, a single presynaptic action potential typically produces only a small depolarization of the postsynaptic cell. The synapse between a single stretch receptor sensory neuron and a quadriceps motor neuron is typical of this situation, and the behavior of this synapse is illustrated in Figure 8-1.

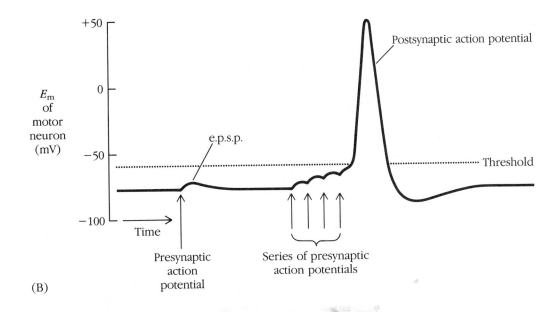

(B)

Temporal and Spatial Summation of Synaptic Potentials

Figure 8-1A shows an experimental arrangement for recording the change in membrane potential of a motor neuron when a single presynaptic sensory neuron is stimulated to fire an action potential. In this case, an intracellular probe is placed inside the sensory fiber so that an action potential can be triggered by passing a depolarizing current into the nerve fiber. Figure 8-1B shows sample results when a single action potential is triggered in the sensory neuron and when four action potentials are triggered in rapid sequence. A single presynaptic action potential produces only a small depolarizing change in the motor neuron's membrane potential. This unitary response to a presynaptic action potential is called an **excitatory postsynaptic potential (e.p.s.p.)**. A single e.p.s.p. is typically much too small to reach threshold, which is usually a depolarization of 10 to 15 mV from resting E_m. Figure 8-2 shows an actual recording of an e.p.s.p. in a motor neuron produced by an action potential in a single sensory neuron. In this experiment, an intracellular electrode was placed inside the sensory fiber to record the presynaptic membrane potential and to inject depolarizing current to elicit an action potential in the presynaptic fiber (upper recording trace). A second intracellular electrode was

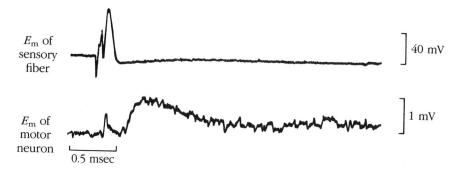

Figure 8-2

Simultaneous intracellular recordings from a single stretch-sensitive sensory nerve fiber and a motor neuron receiving synaptic input from the sensory fiber. An action potential was triggered in the sensory fiber by passing a depolarizing electrical current through the intracellular electrode. After a brief delay, there was a small excitatory postsynaptic potential evoked in the postsynaptic motor neuron. [Unpublished data kindly provided by W. Collins, M. Honig, and L. M. Mendell of the State University of New York, Stony Brook.]

placed in the motor neuron to record the change in membrane potential of the postsynaptic cell, as diagrammed in Figure 8-1A.

If a second action potential arrives at the presynaptic terminal before the postsynaptic effect of a first action potential has disappeared, the second e.p.s.p. will sum with the first to produce a larger peak postsynaptic depolarization. As shown in Figure 8-1B, if a series of action potentials arrives in the presynaptic terminal sufficiently rapidly, it is possible for the individual e.p.s.p.'s to add up to a depolarization that reaches threshold. This kind of summation of sequential postsynaptic effects of an individual presynaptic input is called **temporal summation**. This is an important mechanism by which even a weak excitatory synaptic input can cause a neuron to fire an action potential.

Another way that e.p.s.p.'s can sum to reach threshold is if several presynaptic neurons fire action potentials simultaneously. In reality, a single neuron (such as the motor neuron in Figure 8-1) in the nervous system receives synaptic inputs from hundreds or even thousands of presynaptic neurons. In the patellar reflex, for example, a single quadriceps motor neuron will receive excitatory synaptic connections from many stretch receptor sensory neurons. A simplified picture of this situation is shown in Figure 8-3A. A single action potential in any one of the presynaptic cells produces only a small postsynaptic depolarization (Figure 8-3B). If several presynaptic cells fire simultaneously, however, their postsynaptic effects add up and can reach threshold. This type of summation is termed **spatial summation** to indicate that it occurs among spatially distinct inputs onto a single postsynaptic cell.

In the patellar reflex, both temporal and spatial summation are important in eliciting the reflexive response. In order to produce reflexive contraction of the quadriceps muscle, a tap to the patellar tendon must stretch the muscle sufficiently to fire a number of

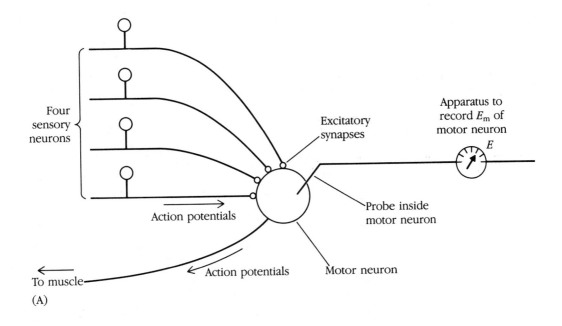

(A)

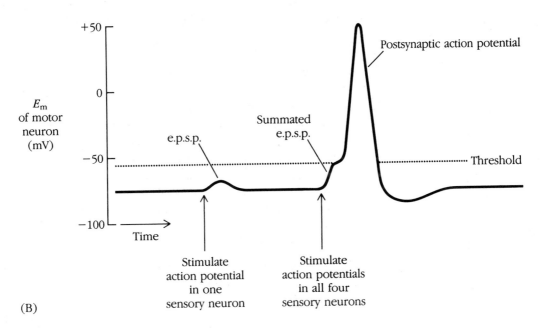

(B)

Figure 8-3
Spatial summation of excitatory inputs to a motor neuron.

action potentials in each of a number of individual sensory neurons.

Some Possible Excitatory Neurotransmitters

The chemical neurotransmitter at the neuromuscular junction has been conclusively demonstrated to be acetylcholine, as discussed in the previous chapter. Unfortunately, the transmitter at most excitatory synapses between neurons in the nervous system is unknown. Many substances are suspected to be excitatory neurotransmitters in the nervous system. Acetylcholine may in fact be the neurotransmitter at some neuron-to-neuron synapses as well as at the neuromuscular junction. The structures of some of these possible excitatory neurotransmitters are shown in Figure 8-4. There is evidence that glutamate may be the excitatory transmitter at the synapse between the sensory and motor neurons in the patellar reflex. The list in Figure 8-4 is by no means exhaustive, and as information about the brain improves it is likely that many new candidates will be added to the list.

Conductance-Decrease e.p.s.p.'s

In most cases, the mechanism by which an excitatory neurotransmitter produces an e.p.s.p. in the postsynaptic cell is the same as that by which acetylcholine depolarizes the muscle at the neuromuscular junction. That is, the transmitter opens channels in the postsynaptic membrane that allow sodium or sodium and potassium ions to cross. This alters the balance of sodium and potassium permeability so that the membrane potential shifts in a positive direction. In terms of the ionic current flowing across the postsynaptic membrane, the excitatory neurotransmitter acts to increase the relative sodium conductance of the postsynaptic membrane, so that the balance between inward sodium current and outward potassium current is struck slightly nearer E_{Na}. We saw in Chapter 4, however, that the membrane potential is controlled by the ratio of sodium to potassium permeability; thus, a depolarization might result either from an increase in sodium permeability or from a decrease in potassium permeability. Indeed, at some synapses, the effect of the transmitter appears to be mediated via a reduction in the potassium conductance of the postsynaptic

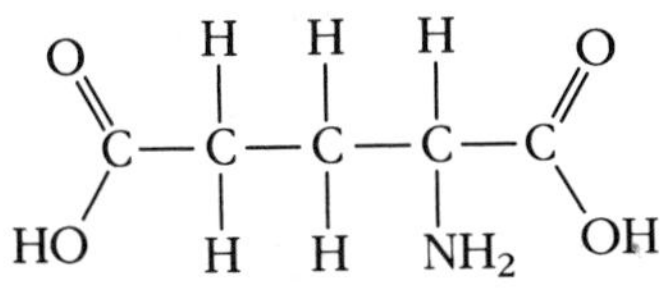

Glutamic acid
(Glutamate)

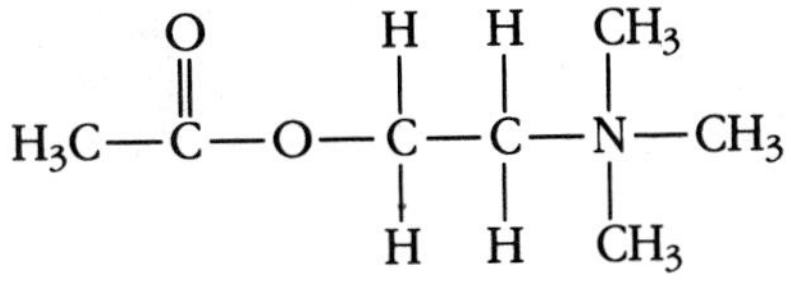

Acetylcholine

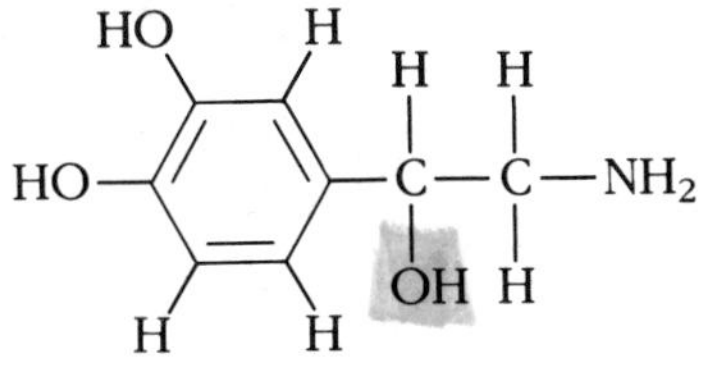

Norepinephrine

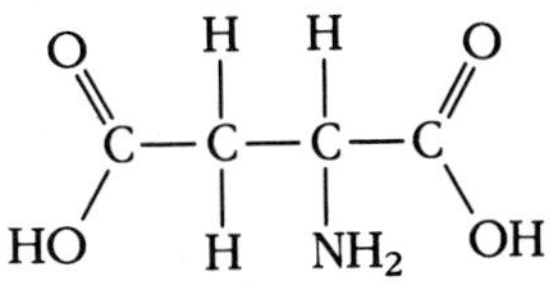

Aspartic acid
(Aspartate)

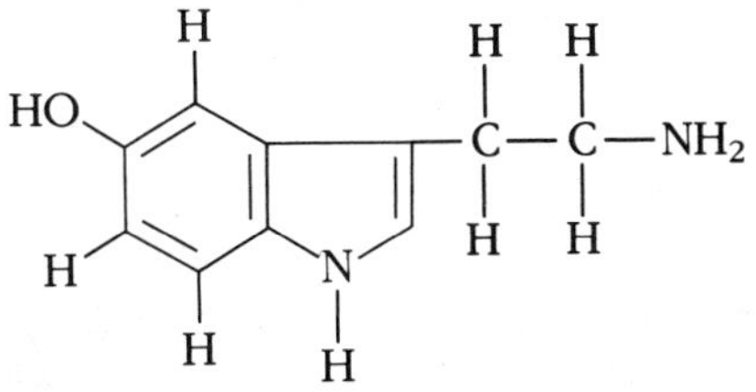

Serotonin
(5-hydroxytryptamine)

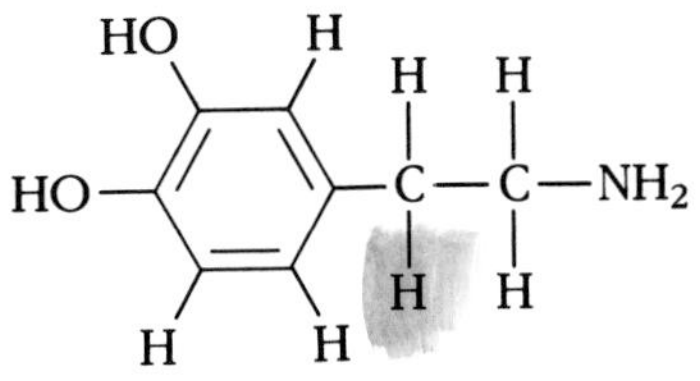

Dopamine

Arg-Pro-Lys-Pro-Gln-Gln-Phe-Phe-Gly-Leu-Met-NH$_2$

Substance P
(a string of 11 amino acids attached by peptide bonds)

Figure 8-4
Structures of some substances likely to be excitatory neurotransmitters in the nervous system.

neuron. For instance, the neurotransmitter serotonin produces a long-lasting depolarization of a certain class of neuron in the central nervous system of a sluglike marine invertebrate, *Aplysia*. This depolarization appears to be mediated by a decrease in the potassium permeability of the neuron; serotonin causes a type of potassium channel in the neuron membrane to close, so that the outward potassium current across the membrane declines and the resting inward sodium current has a greater influence on membrane potential.

Inhibitory Synaptic Transmission

The Sensory Neuron to Antagonist Motor Neuron Synapse in the Patellar Reflex

The diagram for the patellar reflex that we have been considering since Chapter 5 is not really quite complete. As shown in Figure 8-5, muscles other than the quadriceps must be taken into account for a more complete description. There are muscles at the back of the thigh that are attached to the lower leg bones and are responsible for flexion of the knee joint, just as the quadriceps is responsible for extension of the knee joint. These muscles are antagonistic to the action of the quadriceps. The antagonistic muscles have their own stretch-sensitive sensory neurons and motor neurons that are connected in a reflex chain just like those of the quadriceps. That is, a stretch of the muscle stimulates action potentials in the sensory neurons, which make excitatory synapses on the motor neurons that cause contraction of the same muscle.

As we've seen previously, when the patellar tendon is tapped, the quadriceps muscle reflexively contracts, causing the knee joint to extend (the "jerk" of the knee-jerk reflex). The extension of the joint stretches the flexor muscles at the back of the thigh, which should then contract because of the action of their own stretch-reflex mechanism. The resulting flexion of the joint should stretch the quadriceps again and elicit reflexive extension, which should again elicit reflexive flexion, and so on. One might expect, then, that a single tap to the patellar tendon would send the knee joint into a series of oscillations that would continue until muscle exhaustion set in.

Why, then, does tapping the patellar tendon elicit only a single knee jerk? The answer lies in the more elaborate neuronal circuitry diagrammed in Figure 8-5. The nerve fibers of the stretch-sensitive sensory neurons from the quadriceps muscle actually branch profusely when they enter the spinal cord and make synaptic connections with many kinds of neuron in addition to the quadriceps motor neuron. Among these other synaptic connections is an excitatory synapse onto a type of neuron that makes an inhibitory synapse on the motor neurons of the antagonistic muscles. Thus,

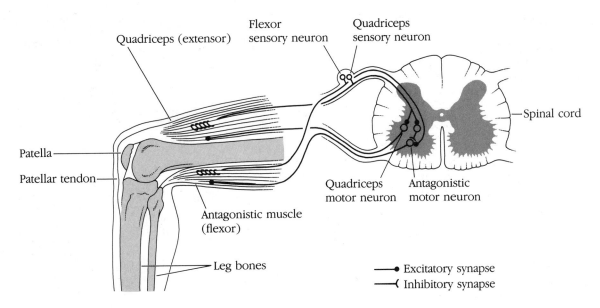

Quadriceps (extensor)
Flexor sensory neuron
Quadriceps sensory neuron
Spinal cord
Patella
Patellar tendon
Quadriceps motor neuron
Antagonistic motor neuron
Antagonistic muscle (flexor)
Leg bones

→● Excitatory synapse
→< Inhibitory synapse

Figure 8-5
A revised diagram of the circuitry involved in stretch reflexes of thigh muscles.

action potentials in quadriceps sensory neurons not only tend to excite quadriceps motor neurons but also tend indirectly to prevent antagonistic motor neurons from being excited by the antagonistic sensory neurons.

Characteristics of Inhibitory Synaptic Transmission

We will now consider some of the properties of postsynaptic responses at an inhibitory synapse and then discuss the underlying mechanisms in the postsynaptic membrane.

Figure 8-6 shows schematically an experimental arrangement to examine the inhibition of the antagonistic motor neuron in the patellar reflex. The membrane potential of the motor neuron is monitored with an intracellular probe while the inhibitory presynaptic neuron is stimulated electrically to fire action potentials. Presynaptically, the basic scheme for release of the neurotransmitter is the same as at other chemical synapses: the depolarization produced by the action potential allows calcium ions to enter the synaptic terminal through voltage-sensitive calcium channels, inducing synaptic vesicles containing the transmitter to fuse with the membrane of the terminal and release their contents. Postsynaptically, however, the effect of the transmitter is very different from that of acetylcholine at the neuromuscular junction.

Figure 8-6

Inhibitory synaptic transmission between two neurons in the circuit of Figure 8-5. An action potential in the presynaptic neuron releases a neurotransmitter that hyperpolarizes the postsynaptic neuron.

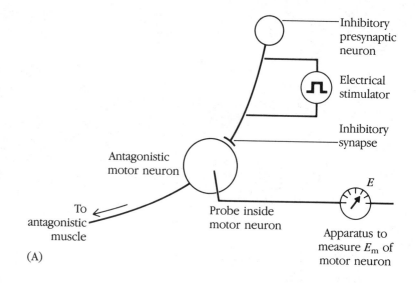

(A)

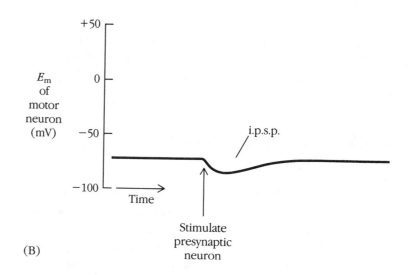

(B)

Sample results are shown in Figure 8-6B. An action potential in the presynaptic cell is followed by a transient *increase* in the postsynaptic membrane potential. When the membrane potential becomes more negative, the cell is said to be **hyperpolarized**. Because a hyperpolarization moves the membrane potential away from the threshold for firing an action potential, it is less likely that an excitatory input will be able to trigger an action potential, and the postsynaptic cell is inhibited. The change in membrane potential of the postsynaptic cell induced by the release of an inhibitory neurotransmitter is called an **inhibitory postsynaptic potential (i.p.s.p.)**.

Mechanism of Inhibition in the Postsynaptic Membrane

A recurring theme in the functioning of cells has been that changes in membrane potential are brought about by changes in ionic permeability of the plasma membrane. The i.p.s.p. is no different in this regard. **When the permeability of the membrane to a particular ion increases, the membrane potential tends to move toward the equilibrium potential for that ion**. We have already seen in Chapter 5 how this sort of mechanism can explain the sequence of potential changes during an action potential. To see how a hyperpolarizing response might result from a change in ionic permeability, consider the situation diagrammed in Figure 8-7.

If potassium permeability of a cell membrane were suddenly increased, the membrane potential would be expected to move toward E_K, which is about -85 mV for a typical mammalian cell (see discussion in Chapters 3 and 4). In this situation, p_{Na}/p_K would be smaller than usual, and the balance between potassium and sodium would be struck closer to E_K. This is similar to the situation during the undershoot at the end of an action potential, when p_{Na}/p_K is transiently smaller than normal. As shown in Figure 8-7B, then, an inhibitory transmitter could hyperpolarize the post-synaptic cell by opening potassium channels in the postsynaptic membrane. As with acetylcholine at the neuromuscular junction, the inhibitory transmitter might act by combining with specific binding sites associated with the gate on the channel. When the binding sites are occupied, the gate controlling movement through the channel opens, and potassium ions can move out of the cell to drive E_m closer to the potassium equilibrium potential.

At most inhibitory synapses, however, it turns out that the transmitter-activated postsynaptic channels are not potassium channels. Instead, most inhibitory neurotransmitters open post-synaptic channels that allow chloride ions to cross the membrane. Recall from Chapters 3 and 4 that chloride pumps in the plasma membrane maintain the chloride equilibrium potential, E_{Cl}, more negative than the resting membrane potential in many nerve cells. That is, the electrical gradient driving chloride out of the cell at the

Figure 8-7
Mechanism by which an inhibitory postsynaptic potential might be produced in a postsynaptic neuron. (A) An increase in potassium permeability would be expected to cause a more negative membrane potential. (B) A neurotransmitter-activated potassium channel in the plasma membrane of a postsynaptic neuron at an inhibitory synapse.

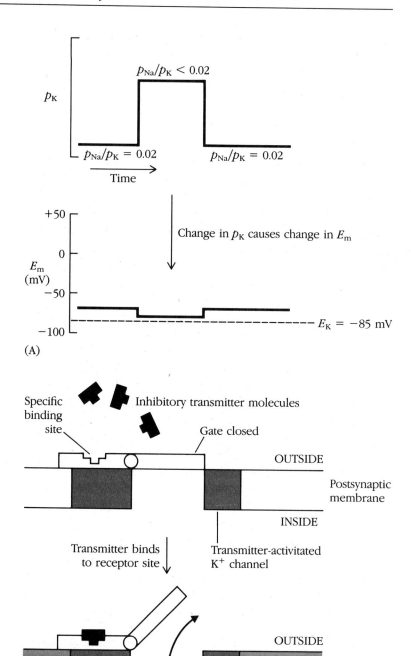

p_K

$p_{Na}/p_K < 0.02$

$p_{Na}/p_K = 0.02$

$p_{Na}/p_K = 0.02$

Time

$+50$

0

E_m (mV)

-50

-100

Change in p_K causes change in E_m

$E_K = -85$ mV

(A)

Specific binding site

Inhibitory transmitter molecules

Gate closed

OUTSIDE

Postsynaptic membrane

INSIDE

Transmitter binds to receptor site

Transmitter-activitated K^+ channel

OUTSIDE

K^+

INSIDE

(B)

normal resting E_m is weaker than the concentration gradient driving chloride into the cell. Thus, if chloride channels open in the postsynaptic membrane, chloride ions will enter the cell, bringing negative charge into the cell and hyperpolarizing the neuron. An i.p.s.p. can result, then, from an increase in the chloride permeability of the postsynaptic cell.

In general terms, inhibition results if a neurotransmitter induces a permeability increase to an ion with an equilibrium potential more negative than the threshold level of membrane potential. This is because ions always move across the membrane in such a direction as to move E_m toward the equilibrium potential for that ion. If that equilibrium potential is more negative than threshold, the ion will always oppose any attempt to reach threshold as soon as the depolarization exceeds the ion's equilibrium potential.

Some Possible Inhibitory Neurotransmitters

The identity of the chemical neurotransmitter at most inhibitory synapses remains unknown. However, several substances are likely candidates, and the structures of some of these are shown in Figure 8-8. Of those shown, GABA and glycine are almost certainly transmitters at certain inhibitory synapses in the nervous system. Note that some of the candidates in Figure 8-8 also appeared in the list of candidate excitatory neurotransmitters (Figure 8-4). This is because a transmitter may have an excitatory effect on one type of neuron but an inhibitory effect on another type of neuron. Whether a substance is excitatory or inhibitory on a given neuron depends on the type of ion channel it opens in the postsynaptic membrane. If the transmitter-activated channel is a sodium or a sodium–potassium channel (as at the neuromuscular junction), the effect will be a depolarization and excitation; if the channel is a chloride or potassium channel, the effect will be hyperpolarization and inhibition. It is even possible for the same neurotransmitter to have opposite effects at two different synapses on the same postsynaptic neuron.

Presynaptic Inhibition

A different type of inhibitory interaction is sometimes observed among neurons in the nervous system. In the inhibition just de-

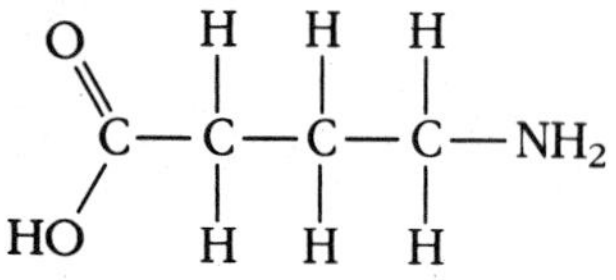

GABA
(γ-aminobutyric acid)

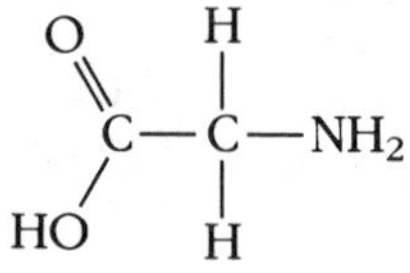

Glycine

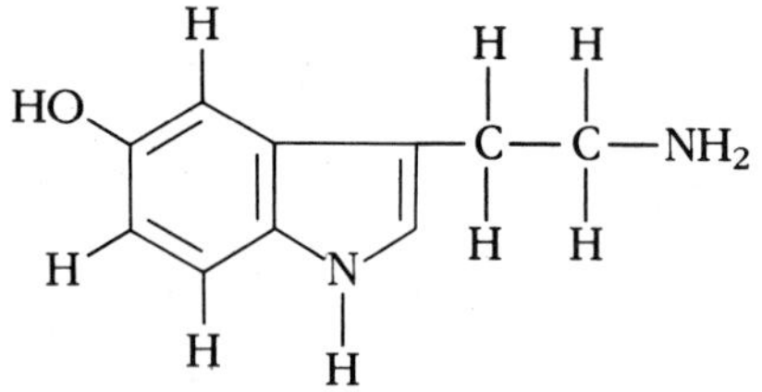

Serotonin
(5-hydroxytryptamine)

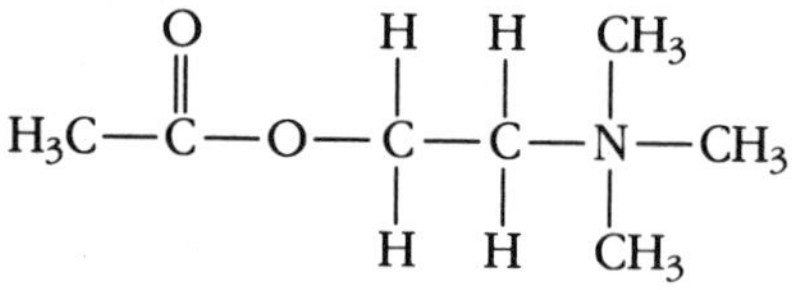

Acetylcholine

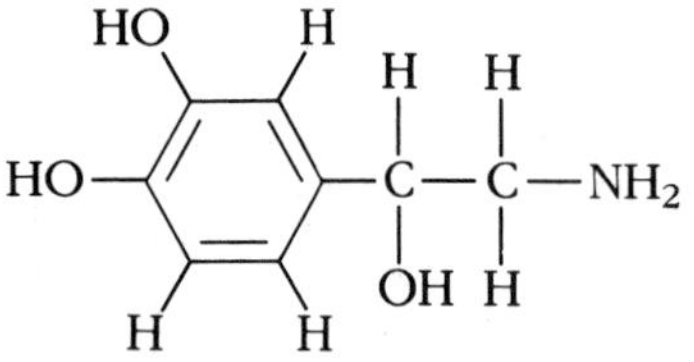

Norepinephrine

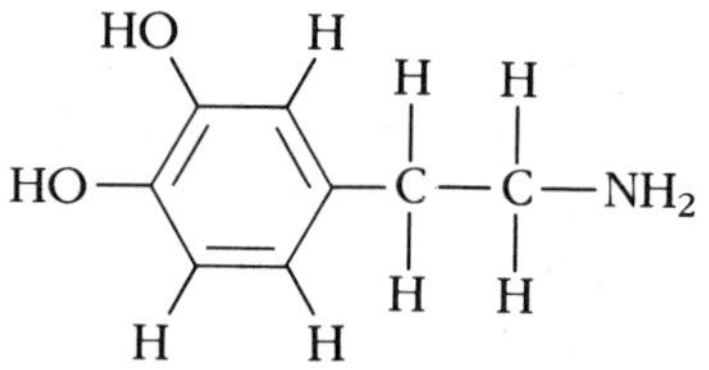

Dopamine

Tyr-Gly-Gly-Phe-Leu
Leucine enkephalin
(a series of five amino acids connected by peptide bonds)

Figure 8-8
Structures of some substances likely to be inhibitory neurotransmitters in the nervous system.

scribed above, the inhibition is accomplished by direct effect of the inhibitory transmitter on the membrane of the postsynaptic cell. In the other type of inhibition, the inhibition is exerted indirectly by affecting the membrane of an excitatory presynaptic terminal. This type of inhibition is called **presynaptic inhibition**, and is illustrated schematically in Figure 8-9. In this situation, the inhibitory terminal acts to reduce the depolarization produced by an action

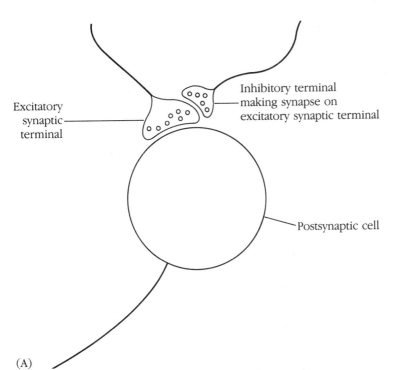

(A)

Figure 8-9
(A) Schematic arrangement for presynaptic inhibition in the nervous system.
(B) Electronmicrograph showing a synapse (Terminal 1) in the vertebrate central nervous system onto an axon (Terminal 2) that in turn makes a synapse onto a third neuronal process (labeled "d" for dendrite). [Photograph reproduced, with permission, from W. O. Wickelgren, *J. Physiol.* 270(1977):89–114.]

Excitatory synaptic terminal

Inhibitory terminal making synapse on excitatory synaptic terminal

Postsynaptic cell

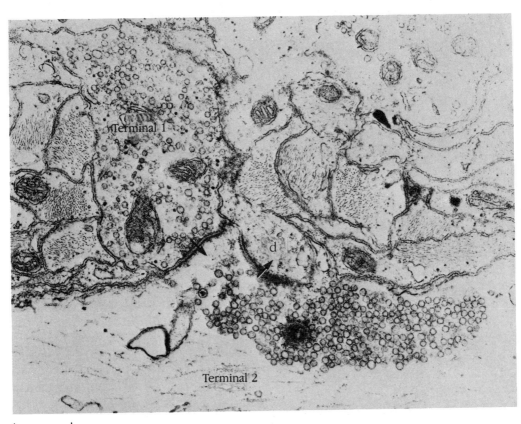

(B) ⊢——⊣ 0.5 μm

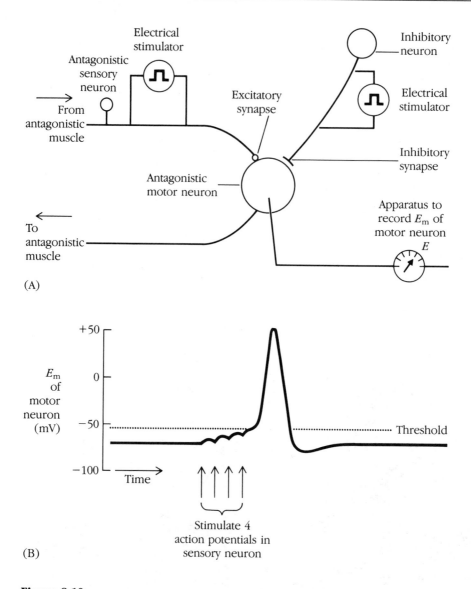

(A)

(B)

Figure 8-10

Integration of excitatory and inhibitory synaptic inputs by a neuron in the nervous system. (A) Experimental arrangement for the measurements shown in (B), (C), and (D). (B) Stimulating the excitatory presynaptic neuron produces a postsynaptic action potential if there is sufficient temporal summation to reach threshold. (C) Stimulating the inhibitory presynaptic neuron prevents the excitatory inputs in (B) from reaching threshold. (D) The inhibitory effect can be overcome by increasing the amount of excitatory stimulation.

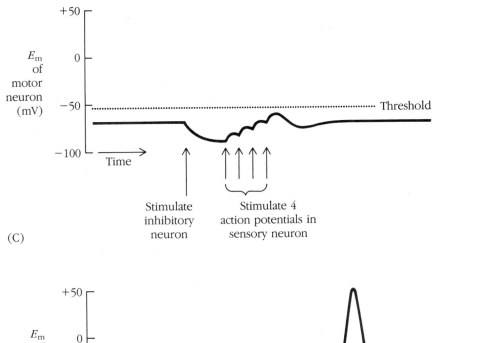

(C)

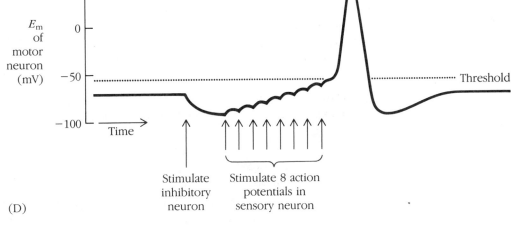

(D)

potential in the excitatory presynaptic terminal that makes a direct connection onto the postsynaptic cell. A smaller depolarization opens fewer voltage-sensitive calcium channels, so that less calcium enters the excitatory terminal. This in turn causes fewer synaptic vesicles to fuse with the terminal membrane. Thus, when the inhibitory terminal is active, the excitatory terminal releases less excitatory transmitter onto the postsynaptic neuron. The electron micrograph of Figure 8-9B shows a synapse being made onto a nerve fiber at a location where that fiber in turn makes a synapse onto another postsynaptic neuron. This kind of arrangement might underlie the presynaptic inhibition of the type just described.

Neuronal Integration

In the nervous system, neurons receive both excitatory and inhibitory synaptic inputs. The decision of a postsynaptic neuron to fire an action potential is determined by only one factor: whether or not the threshold level of membrane potential has been reached. This is determined at any instant by the sum of all existing excitatory and inhibitory synaptic potentials. This can be seen by looking once again at the antagonistic motor neuron in the patellar reflex, as shown in Figure 8-10.

When the sensory neuron from the antagonistic muscle is stimulated to fire action potentials, e.p.s.p.'s result in the motor neuron, and an action potential results if there is sufficient temporal summation among the e.p.s.p.'s. If the inhibitory neuron is stimulated at the same time, however, the same series of excitatory inputs might not be able to reach threshold. This inhibitory effect could be overcome by increasing the strength of the excitatory input, either by increasing the number of presynaptic action potentials in the sensory neuron or by increasing the number of sensory neurons activated. It is the balance between the excitatory and inhibitory inputs that governs whether or not a postsynaptic action potential is generated.

The process of summing up, or integrating, all the synaptic inputs onto a neuron is called **neuronal integration**, and it underlies all the information transfer and processing capabilities of the nervous system. The information processing capacity of a single neuron is considerable in that a typical neuron receives hundreds or thousands of synapses from hundreds or thousands of other neurons and makes synaptic connnections itself with an equal number of postsynaptic neurons. This capacity is increased still further by the fact that synaptic inputs to a cell have widely varying weights: some synapses produce large changes in postsynaptic membrane potential, while others cause only tiny changes. Further, the weight given a particular input might vary with time, as in the case of presynaptic inhibition. A network of some 10^{10} of these sophisticated units, like the human brain, has staggering information processing ability.

Summary

Chemical synapses between neurons in the nervous system are similar to the synapse at the neuromuscular junction in the following ways:

1. Neurotransmitter is stored in the synaptic terminal in membrane-bound synaptic vesicles.

2. Influx of external calcium ions into the terminal is a necessary step in the release of neurotransmitter.

3. Fusion of the synaptic vesicle with the plasma membrane of the terminal is the mechanism of transmitter release.

4. Neurotransmitter molecules combine with specific postsynaptic receptors and open ion channels in the postsynaptic membrane.

Nervous system synapses are different from the neuromuscular junction in the following ways:

1. At most synapses, a single presynaptic action potential produces only a small change in postsynaptic membrane potential. At the neuromuscular junction, a single presynaptic action potential produces a large depolarization of the muscle cell and triggers a postsynaptic action potential.

2. Synapses between neurons can be either excitatory or inhibitory.

3. Acetylcholine is probably the neurotransmitter at some synapses between neurons, but a host of other substances are also probably neurotransmitters in the nervous system.

4. A skeletal muscle cell receives synaptic input from only one neuron, a single motor neuron. A neuron in the nervous system may receive synaptic connections from thousands of different neurons. The output of a neuron depends on the integration of all the inhibitory and excitatory inputs active at a given instant.

PART III

CELLULAR PHYSIOLOGY OF MUSCLE CELLS

This section will be concerned with the second major type of excitable cell: muscle cells. These cells are specialized to produce movement when electrically stimulated, and their actions are the most obvious external manifestation of the activities of the nervous system. In Chapter 7, the point of interaction between the nervous and muscular systems—the neuromuscular junction—was the central focus for the discussion of chemical synaptic transmission. In the first chapter of Part III, Chapter 9, we will return to the neuromuscular junction to examine the sequence of events that link an action potential in the postsynaptic muscle cell to mechanical contraction. This linkage is the process called **excitation–contraction coupling**, and the explanation of this process in terms of underlying molecular mechanisms stands as one of the major accomplishments of modern cellular physiology. In Chapter 10 we will discuss some of the characteristics of contractions of whole skeletal muscles and how the strength of contraction is controlled by the nervous system. Chapter 11 will consider some of the important electrical differences between the cells of skeletal muscles and the muscle cells of the heart; these electrical differences underlie the ability of the heart as an organ to produce the rhythmic, coordinated contractions necessary to pump the blood through the body.

9

Excitation–Contraction Coupling in Skeletal Muscle

The final stage of the patellar reflex (see Chapter 5) is the contraction of the quadriceps muscle brought about by the activity of the motor neurons making excitatory synaptic connections with that muscle. The arrival of an action potential in the synaptic terminal of the presynaptic motor neuron causes release of the chemical neurotransmitter, acetylcholine. The acetylcholine in turn depolarizes the end-plate region of the postsynaptic muscle cell, initiating an action potential in the muscle cell. This action potential propagates along the long, thin muscle cell just as the neuron action potential propagates along nerve fibers. The muscle action potential serves as the trigger for contraction of the muscle cell. This chapter will examine the events that intervene between the occurrence of the action potential in the plasma membrane of the muscle cell and the activation of the contraction: the process of excitation–contraction coupling.

To begin, it will be useful to examine the structure of the muscle cells at the level of both the light and electron microscopes. We will then consider the molecular makeup of the contractile apparatus and discuss the biochemical mechanisms that control the action of that apparatus. The chapter will conclude with a discussion of the current state of knowledge about the interaction between the muscle action potential and those control mechanisms.

The Three Types of Muscle

There are two general classes of muscle in the body: striated and smooth. Both are named for the characteristic appearance of the individual cells making up the muscle tissue when viewed through a microscope; striated muscle cells (see Figure 9-1) have closely spaced, crosswise stripes (striations), but smooth muscle cells do not and thus appear smooth. Smooth muscle is found in the gut, blood vessels, the uterus, and other locations in which contractions are usually slow and maintained. The muscles that move and support the skeletal framework of the body—the skeletal muscles—are made up of striated muscle cells. This chapter will focus on the structure and properties of skeletal muscle cells.

The cells that make up the muscle of the heart are also striated, like skeletal muscle. Because the membranes of cardiac muscle cells are electrically quite different from those of skeletal muscle cells, cardiac muscle is usually regarded as a distinct class of muscle in its own right. The characteristics of cardiac muscle will be discussed in Chapter 11.

Structure of Skeletal Muscle

Figure 9-1 shows the structure of a typical mammalian skeletal muscle at progressively greater magnification. To the naked eye, an intact muscle appears to be vaguely striped longitudinally, as in Figure 9-1A. Upon closer inspection, the muscle is made up of bundles of individual cells: the muscle cells or muscle fibers (see Figure 9-1B). In mammalian muscle, the individual cells are about 50 μm in diameter and are as long as the whole muscle. Thus, muscle cells are typically long, thin fibers. Because the end-plate region typically extends over only a few microns, a rapidly propagating action potential is required in skeletal muscle cells to pass along the motor neuron's command to contract.

Individual muscle cells consist of bundles of still smaller fibers called **myofibrils**. Thus, the plasma membrane of a single muscle cell encloses many myofibrils. At the level of the myofibrils, the structural basis of the crosswise striations of skeletal muscle cells begins to become apparent. As shown in Figure 9-1C, myofibrils exhibit a repeating pattern of crosswise light and dark stripes. Early anatomists who first described these stripes had no idea of their function and gave them the cryptic names A band, I band, and Z

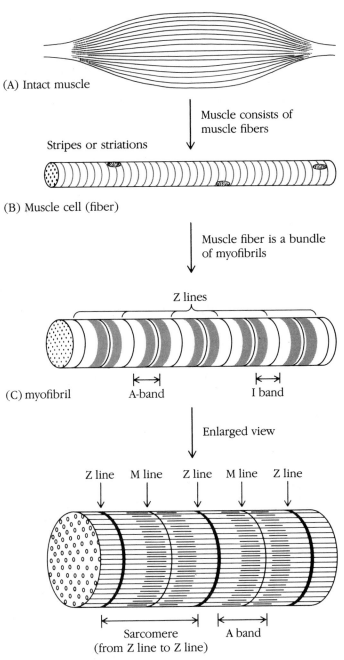

(A) Intact muscle

Muscle consists of
muscle fibers

Stripes or striations

(B) Muscle cell (fiber)

Muscle fiber is a bundle
of myofibrils

Z lines

(C) myofibril   A-band   I band

Enlarged view

Z line   M line   Z line   M line   Z line

Sarcomere   A band
(from Z line to Z line)

(D) Two sarcomeres

Figure 9-1
Microscopic structure of a
skeletal muscle.

line. The I band is a predominantly light region with the dark Z line at its center, while the A band is a darker region separating two I bands of the repeating pattern. At still higher magnification, the A band can be seen to have its own internal structure (see Figure 9-1D); two darker areas at the outer edges of the A band are separated by a lighter region with a faint dark line, called the M line, at the center. The basic unit of the repeating striation pattern of a myofibril is called a **sarcomere**, and it is defined as extending from one Z line to the next—that is, from the center of one I band to the center of the next I band. Thus, Figure 9-1D shows two complete sarcomeres along the length of a single myofibril.

Changes in Striation Pattern on Contraction

All of these myofibril striation patterns would be rather uninteresting, except to anatomists, if it weren't for the important observation that the relation among the stripes changes when a muscle cell contracts. This change can best be appreciated at the level of the electron microscope, as shown diagrammatically in Figure 9-2. Through the electron microscope, a myofibril can be seen to consist of two kinds of longitudinally oriented filaments, called thick and thin filaments. Both the thick and thin filaments are arrayed in parallel groups. As shown in Figure 9-2, the Z line corresponds to the position where the thin filaments of one sarcomere join onto those of the neighboring sarcomere and where cross connections are made among the parallel thin filaments. The thick filaments within a sarcomere are joined to each other at the M line. It is clear from comparing Figure 9-2 with Figure 9-1D that the lighter I band corresponds to the region occupied only by thin filaments, and the darker A band corresponds to the spatial extent of the thick filaments. The darker regions at the two edges of the A band correspond to the region of overlap of the thick and thin filaments. The thick filaments bear thin fibers that appear to link to the thin filaments in the region of overlap, forming cross bridges between the thick and thin filaments.

Upon contraction, the length of each sarcomere shortens—that is, the distance between successive Z lines diminishes. However, the width of the A band is unaffected by contraction; thus, only the I band becomes thinner during a contraction. In terms of the thick and thin filaments, this observation can be explained by the sliding filament hypothesis, which is illustrated in Figure 9-2B. Neither the

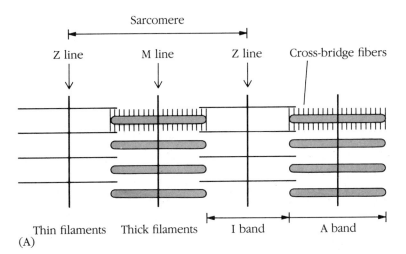

Sarcomere

Z line   M line   Z line   Cross-bridge fibers

Thin filaments   Thick filaments   I band   A band

(A)

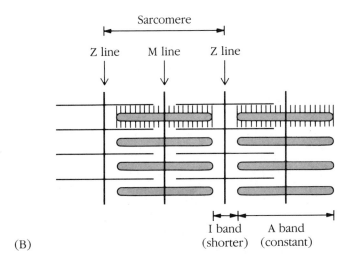

Sarcomere

Z line   M line   Z line

I band   A band
(shorter)   (constant)

(B)

Figure 9-2
Schematic representation of the relationships between thick and thin filaments of a myofibril in (A) a relaxed muscle and (B) a contracted muscle.

thick nor the thin filaments change in length during a contraction; rather, shortening occurs because the filaments slide with respect to one another, so that the region of overlap increases. In order to understand how the sliding occurs, it will be necessary to examine the molecular makeup of the thick and thin filaments.

Molecular Composition of Filaments

The thick filaments are aggregates of a protein called myosin, which consists of a long fibrous "tail" connected to a globular "head" region, as shown schematically in Figure 9-3. The fibrous

Figure 9-3

Diagram of the structure of a single molecule of the thick filament protein, myosin. The fibrous tail is connected to the globular head region via a flexible connection point. The globular head includes a region that can bind and split a molecule of ATP.

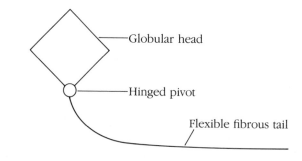

Figure 9-4

Hypothetical structure of a thick filament. The fibrous tails of individual myosin molecules polymerize to form the backbone of the filament. The globular heads radiate out perpendicular to the long axis of the filament to form the cross-bridges to the thin filament. The myosin molecules reverse orientation at the M line.

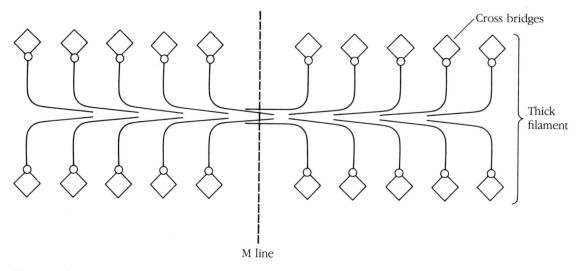

tails tend to aggregate into long filaments, with the heads projecting off to the side. Figure 9-4 shows a generally accepted view of how myosin molecules are arranged in the thick filaments of a sarcomere. The aggregated tails form the backbone of the thick filament, and the globular heads form the cross bridges that connect with adjacent thin filaments.

The globular head of myosin contains a region that can bind adenosine triphosphate (ATP) and split one of the high-energy phosphate bonds of the ATP, releasing the stored chemical energy. That is, myosin acts as an ATPase. The energy provided by the ATP is transferred to the myosin molecule, which is transformed into an "energized" state. This sequence can be summarized as follows:

$$\text{Myosin} + \text{ATP} \rightarrow \text{Myosin·ATP} \rightarrow \text{Energized Myosin·ADP} + P_i$$

Here, the dot indicates that two molecules are bound together, as in an enzyme-substrate complex. To make a mechanical analogy,

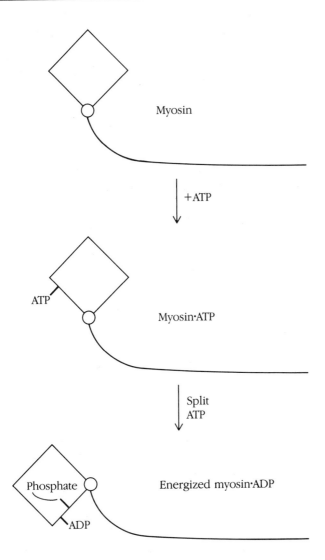

Figure 9-5
A molecule of myosin can bind a molecule of ATP. Energy released by splitting a high-energy phosphate bond of the ATP is used to form an energized form of myosin. The transition from resting to energized state involves rotation of the globular head about its flexible attachment site to the fibrous tail.

the globular head behaves as though it is attached to the fibrous tail at a hinged connection point. The energy released by ATP causes the head to pivot about the hinge into the energized state, as drawn schematically in Figure 9-5. To continue with mechanical analogies, this can be thought of as cocking the spring-loaded hammer of a cap pistol. As we will see shortly, the energy stored in this energized form of myosin is the energy that fuels the sliding of the filaments during contraction.

The thin filaments within a myofibril are largely made up of the protein actin. Thin filaments also contain two other kinds of protein molecule called troponin and tropomyosin, whose roles in con-

Figure 9-6

Schematic representation of the interaction between energized myosin and actin according to the sliding filament hypothesis. Energized myosin binds to a specific binding site on actin, and the stored energy is released. The resulting relaxation of the myosin molecule entails rotation about the flexible attachment site to the rest of the thick filament. The rotation induces longitudinal sliding of the filaments. The fibrous tail connecting the myosin molecule to the rest of the thick filament is assumed to be "springy," so that the globular head can reach across to the cross-bridge attachment site on the thin filament.

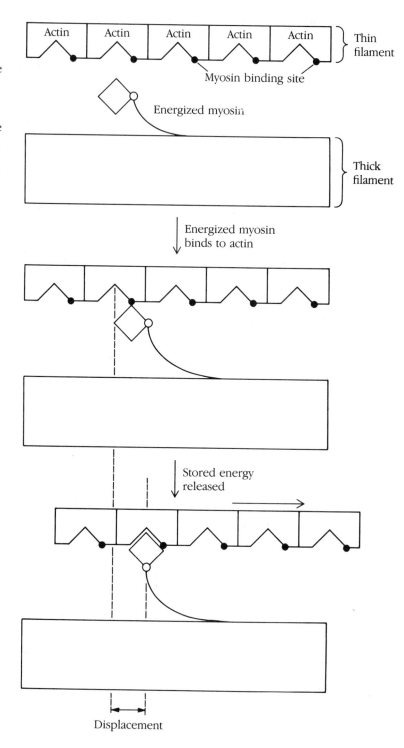

traction will be discussed a bit later; for the present we will concentrate on the actin molecules. Actin is a globular protein that
polymerizes to form long chains; thus, the thin filament can be
thought of as a long string of actin molecules, like a pearl necklace.
(Actually, each thin filament consists of two actin chains entwined
about each other in a helix, but for a conceptual understanding of
the sliding filament hypothesis it is not necessary to keep this in
mind.) Each actin molecule in the chain contains a binding site that
can combine with a specific site on the globular head of a myosin
molecule. This is the site of attachment of the cross bridge on the
thin filament.

Interaction Between Myosin and Actin

When actin combines with energized myosin, the stored energy in
the myosin molecule is released. This causes the myosin molecule
to return to its resting state, and the globular head pivots about its
hinged attachment point to the thick filament. The pivoting motion
requires that the thick and thin filaments move longitudinally with
respect to each other. This mechanical analog is illustrated schematically in Figure 9-6. The exact nature of the chemical changes in
a myosin molecule during the transition from resting to energized
state and back is unknown at present; the sliding filament hypothesis, however, requires that there be some chemical equivalent of
the hinged arrangement shown in Figure 9-6.

How is the bond between actin and myosin broken so that a new
cycle of sliding can be initiated? In the scheme presented so far,
each myosin molecule on the thick filament could interact only one
time with an actin molecule on the thin filament, and the total
excursion of sliding would be restricted to that produced by a
single pivoting of the globular head. In order to produce the large
movements of the filaments that actually occur, it is necessary that
the attachment of the cross bridges be broken so that the cycle of
myosin energization, binding to actin, and movement can be repeated. The full cycle that allows this to occur is summarized in
Figure 9-7. When energized myosin binds to actin and releases its
stored energy, the ADP bound to the ATPase site of the globular
head is released. This allows a new molecule of ATP to bind to the
myosin. When this happens, the bond between actin and myosin is
broken, possibly because of structural changes in the globular head
induced by the interaction between ATP and myosin. The new ATP

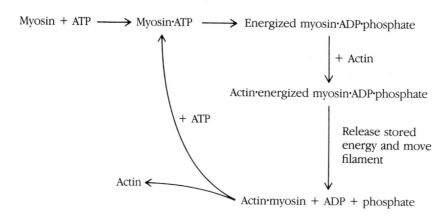

Figure 9-7
The cycle of cross-bridge formation and dissociation between myosin and actin during filament sliding.

molecule can then be split by myosin to regenerate the energized form, which is then free to interact with another actin molecule on the chain making up the thin filament. Note that there are two roles for ATP in this scheme: to provide the energy to "cock" myosin for movement, and to break the interaction between actin and myosin after movement has occurred. If there is no ATP present, actin and myosin get stuck together and a rigid muscle results (as in rigor mortis).

Each of the many myosin heads on an individual thick filament independently goes through repetitive cycles of energization by ATP, binding to actin, releasing stored energy to produce sliding, and detachment from actin. Each cycle results in the splitting of one molecule of ATP to ADP and inorganic phosphate. Note from Figure 9-4 that the orientation of the myosin heads reverses at the midpoint of the thick filament, the M line. This is the proper orientation to pull both Z lines at the boundary of a sarcomere toward the center (see Figure 9-8). The thin filaments attached to the left Z line will be pulled to the right by the cyclical pivoting of the myosin cross bridges. Similarly, the thin filaments attached to the right Z line will be pulled to the left. Thus, each sarcomere in each myofibril shortens, and the whole muscle shortens.

Regulation of Contraction

In the scheme summarized in Figure 9-7, there is no mechanism to control the interaction between actin and myosin. That is, as long as ATP is present, we would expect every muscle in the body to be in a perpetual state of maximum contraction. This section will examine the molecular mechanisms that prevent the interaction of actin with

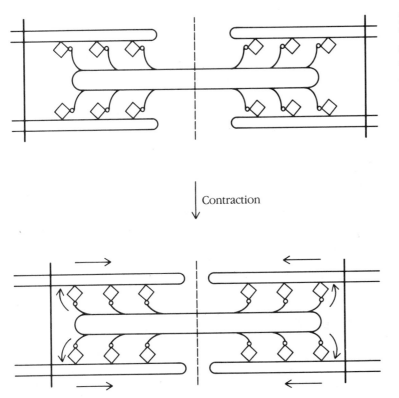

Contraction

Figure 9-8
The mechanism of sarcomere shortening during contraction according to the sliding filament hypothesis. For clarity, the myosin heads are shown acting in concert, although in reality they behave independently.

myosin except when a contraction is triggered by an action potential in the muscle cell membrane.

Recall that thin filaments also contain the proteins troponin and tropomyosin. These proteins are responsible for regulating the interaction between individual myosin and actin molecules in the thick and thin filaments. The regulatory scheme is summarized by the diagrams in Figure 9-9. In the resting muscle, tropomyosin is in a position on the thin filament that allows it to effectively cover the myosin binding site on actin. Myosin's access to the binding site is blocked by the tropomyosin. The position of tropomyosin on the actin polymer is in turn regulated by troponin. In the resting state, troponin locks tropomyosin in the blocking postion. Thus, tropomyosin acts like a trap door covering the myosin binding site, and troponin acts like a lock to keep the door from opening.

What is the trigger that causes tropomyosin to reveal the myosin binding sites on actin? The signal that initiates contraction is the binding of calcium ions to troponin. Each troponin molecule contains a specific binding site for a single calcium ion. Normally, the concentration of calcium inside the cell is very low, and the binding

Figure 9-9
Regulation of the interaction between actin and myosin by calcium ions, troponin, and tropomyosin. In the absence of calcium ions, access to the myosin binding site of actin is prevented by tropomyosin. When calcium binds to troponin, the binding site is accessible and the sliding interaction is allowed.

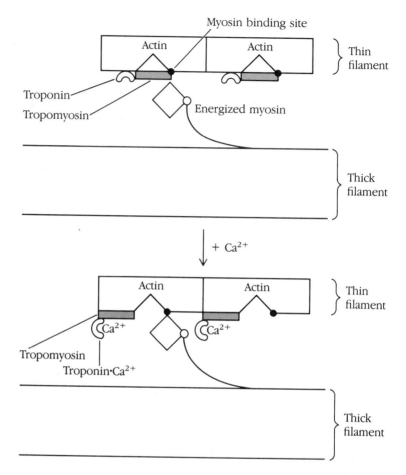

site is not occupied. It is in this state that troponin locks tropomyosin in the blocking position. When an action potential occurs in the muscle cell plasma membrane, however, the concentration of calcium ions in the intracellular fluid rises dramatically, and calcium binds to troponin. When this happens, there is a structural change in the troponin molecule, and the interaction between troponin and tropomyosin is altered in such a way that tropomyosin uncovers the myosin binding site on actin. The cycle of events depicted in Figure 9-7 is then allowed to occur, and the filaments slide past each other.

The Sarcoplasmic Reticulum

Where does the calcium come from to trigger the interaction of actin and myosin underlying the sliding of the filaments? Recall from Chapter 7 that a rise in internal calcium is also responsible for the release of chemical transmitter during synaptic transmission, and that in that case the calcium enters the cell from the ECF through voltage-sensitive calcium channels in the plasma membrane. In the case of skeletal muscle, however, the calcium does not come from outside the cell; rather, the calcium is injected into the intracellular fluid from a separate intracellular compartment called the sarcoplasmic reticulum. The sarcoplasmic reticulum is an intracellular sack that surrounds the myofibrils of a muscle cell. This sack forms a separate intracellular compartment, bounded by its own membrane that is not continuous with the plasma membrane of the muscle cell.

The concentration of calcium ions inside the sarcoplasmic reticulum is much higher than it is in the rest of the space inside the cell; in fact, it is probably higher than the concentration of calcium in the ECF. This accumulation of calcium inside the sarcoplasmic reticulum is accomplished by a calcium pump in its membrane. Like the sodium pump of the plasma membrane, this calcium pump uses metabolic energy in the form of ATP to transport calcium ions across the membrane against a large concentration gradient; in this case, the pump moves calcium ions into the sarcoplasmic reticulum. To initiate a contraction, a puff of calcium ions is released from the sarcoplasmic reticulum, which is strategically located surrounding the contractile apparatus of the myofibrils. The action of the released calcium is terminated as the ions are pumped back into the sarcoplasmic reticulum by the calcium pump. Here, then, is a third role for ATP in the contraction process: ATP, as the fuel for the calcium pump, is responsible for terminating a contraction as well as for energizing myosin and breaking the bond between actin and myosin.

The Transverse Tubule System

How does an action potential in the plasma membrane of a muscle cell trigger release of calcium from the sarcoplasmic reticulum? It has been demonstrated that the crucial aspect of the action potential in triggering contraction is depolarization of the plasma membrane; any manipulation that produces depolarization can cause contraction. How can depolarization of the plasma membrane at the outer surface of the cell affect sarcoplasmic reticulum surrounding myofibrils deep in the interior of the muscle cell? The

Figure 9-10
The transverse tubules are invaginations of the plasma membrane of the muscle fiber. Depolarization during an action potential can spread along the transverse tubules to the interior of the fiber. The triad is a region where the membranes of the tubule and the sarcoplasmic reticulum come into close proximity and presumably represents the area where depolarization acts to release calcium from the sarcoplasmic reticulum.

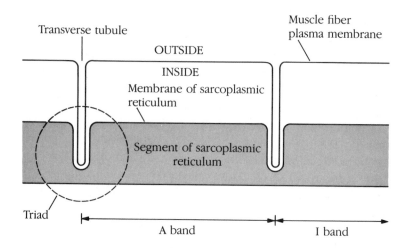

answer is that there are periodic infoldings, or invaginations, of the plasma membrane that extend into the depths of the muscle fiber. These invaginations, called the transverse tubule system, are illustrated in Figure 9-10. The long fingers of plasma membrane projecting into the cell provide a path for depolarization resulting from an action potential in the surface membrane to influence events in the interior of the cell.

In most animal species, the transverse tubules occur just at the boundary between the A band and the I band. At locations where the transverse tubule encounters the sarcoplasmic reticulum, the membranes come into close apposition to form a structure called a triad. This is presumably the point of interaction between the depolarizing signal and the membrane of the sarcoplasmic reticulum. Note, however, that although the membranes are close together at a triad, they do not touch. Because the membranes are not in continuity, the depolarization produced during the action potential cannot spread directly to the sarcoplasmic reticulum. It is not yet clear what signal links depolarization of the transverse tubule to calcium release by the sarcoplasmic reticulum.

Summary

The sequence of events leading to contraction of a skeletal muscle fiber following stimulation of its motor neuron can be summarized as follows:

1. Acetylcholine released from the presynaptic terminal depolarizes the end-plate region of the muscle fiber.

2. The depolarization initiates an all-or-none action potential in the muscle fiber, and the action potential propagates along the entire length of the fiber.

3. Depolarization produced by the action potential spreads to the interior of the fiber along the transverse tubule system.

4. Depolarization of the transverse tubules causes release of calcium ions by the sarcoplasmic reticulum.

5. Released calcium ions bind to troponin molecules on the thin filaments.

6. When calcium combines with troponin, tropomyosin uncovers the myosin binding site of actin.

7. Globular heads of myosin molecules, which have been energized by splitting a high energy phosphate bond of ATP, are then free to bind to actin.

8. The stored energy of the activated myosin is released to propel the thick and thin filaments past each other. The spent ADP is released from myosin at this point.

9. A new ATP binds to myosin, releasing it attachment to the actin molecule.

10. The new ATP is split to reenergize myosin and return the contraction cycle to step 7 above.

11. Contraction is maintained as long as internal calcium concentration is elevated. The calcium concentration falls as calcium ions are taken back into the sarcoplasmic reticulum via an ATP-dependent calcium pump.

Contraction of Whole Muscle

In Chapters 7 and 9, we were concerned with the physiology of muscle at the level of single muscle fibers. We have come to some understanding of the mechanisms involved in the linkage between an action potential in a presynaptic motor neuron and the contraction of the postsynaptic muscle cell. We saw that the motor neuron depolarizes the muscle fiber, causing an action potential that, in turn, triggers the all-or-none contraction of the fiber. In this chapter, we will step back a bit in perspective and look at the functioning of a skeletal muscle as a whole. We will consider how the all-or-none twitches of single muscle fibers are integrated into the smooth, graded movements we know our muscles are capable of making.

The Motor Unit

A single motor neuron makes synaptic contact with a number of muscle fibers. The actual number varies considerably from one muscle to another and from one motor neuron to another within the same muscle; a single motor neuron may contact as few as 10 to 20 muscle fibers or more than 1000. However, in mammals, a single muscle fiber normally receives synaptic contact from only one motor neuron. Therefore, a single motor neuron and the muscle fibers to which it is connected form a basic unit of motor organization called the **motor unit**. A schematic diagram of the organization of a motor unit is shown in Figure 10-1. Recall from Chapter 7 that the synapse between a motor neuron and a muscle

Figure 10-1

Schematic illustration of the innervation of a small number of muscle fibers in a muscle. The shaded muscle fibers form part of the motor unit of motor neuron 1 and the unshaded fibers form part of the motor unit of motor neuron 2.

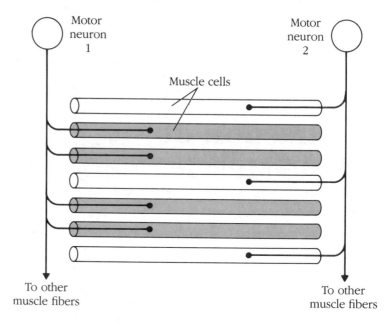

fiber is a one-for-one synapse—that is, a single presynaptic action potential produces a single postsynaptic action potential and hence a single twitch of the muscle cell. This means, then, that all the muscle cells in a motor unit contract together and that the fundamental unit of contraction of the whole muscle will not be the contraction of a single muscle fiber, but the contraction produced by all the muscle cells in a motor unit.

Gradation in the overall strength with which a particular muscle contracts is under control of the nervous system. There are two basic ways the nervous system uses to accomplish this task: (1) variation in the total number of motor neurons activated, and hence in the total number of motor units contracting; and (2) variation in the frequency of action potentials in the motor neuron of a single motor unit. The greater the number of motor units activated, the greater the strength of contraction; similarly, within limits, the greater the rate of action potentials within a motor unit, the greater the strength of the resulting summed contraction. We will consider each of these mechanisms in turn below.

The Mechanics of Contraction

When the nerve controlling a muscle is stimulated, the resulting action potentials in the muscle fibers set up the sliding interaction between the filaments of the individual myofibrils in the muscle. This sliding generates a force that tends to make the muscle fibers, and therefore the muscle as a whole, shorten. Whether or not the muscle actually shortens, however, depends on the load attached to the muscle. While we might attempt to order the muscles in our arms to lift an automobile, it is unlikely that the muscles would be able to shorten against such a load. The force developed in an activated muscle is called the muscle **tension**, and only if the tension is great enough to equal the weight of the load will the muscle shorten and lift the load.

We can distinguish between two kinds of response to activation of a muscle. If the muscle tension is less than the load, the contraction is said to be **isometric** ("same length") because the length of the muscle does not change even though the tension increases. That is, the force exerted on the load by the muscle is not sufficient to move the load, so the muscle cannot shorten. An isometric contraction is diagrammed in Figure 10-2A. In the figure, an isolated muscle is attached to a load it cannot lift. When the muscle is activated, the resulting tension is registered by a strain gauge that measures the miniscule flexing of the rigid strut to which the muscle is attached. A single activation of the muscle triggers a transient increase in tension lasting typically about 0.1 sec. You can easily feel the tension developed in an isometric contraction by placing your palms together with your arms flexed in front of your chest and pushing with both hands, one against the other.

If the tension is great enough to overcome the weight of the load, the contraction is said to be **isotonic** ("same tension") because the tension remains constant once it reaches the level necessary to move the load. This situation is diagrammed in Figure 10-2B. The strain gauge again records the increase in tension, as with the isometric contraction. When the tension reaches the level necessary to lift the load, it levels off and the muscle begins to shorten as the load is lifted. During the change in muscle length, the tension remains constant and equal to the weight of the load. This is because it is this weight—hanging from the muscle and support strut—that determines the flexing measured by the strain gauge. Thus, while the muscle is changing length the contraction is isotonic. In an isotonic contraction, the force developed by the sliding filaments in the myofibrils making up the muscle produces work in the form of moving the load through space.

One difference between isometric and isotonic contractions can be seen in the different delays between muscle activation and the

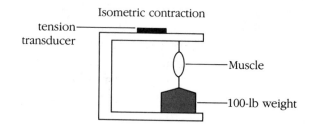

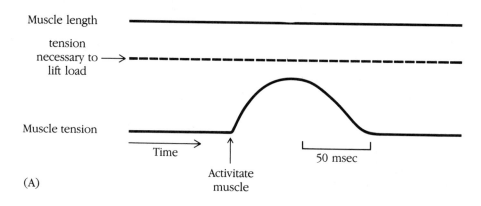

(A)

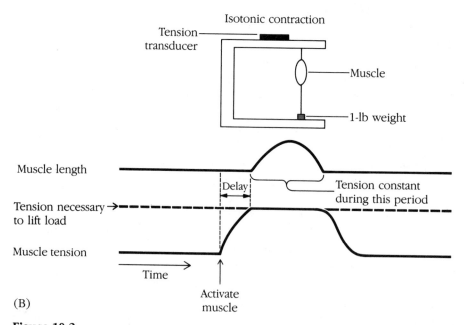

(B)

Figure 10-2
Measures of muscle length and muscle tension during (A) isometric and (B) isotonic contractions. At the upward arrow, the nerve innervating the muscle is stimulated, causing activation of the muscle fibers.

occurrence of a measurable change in either muscle tension (isometric) or muscle length (isotonic). The tension begins to rise within a few milliseconds, the time required for the effect of the excitation–contraction process discussed in Chapter 9 to take hold. However, if muscle length is measured instead there is a pronounced delay between activation of the muscle and beginning of shortening. This delay is the time required for the tension to rise to the point where the load is lifted, which will depend on the size of the load. Thus, with light loads the shortening begins quickly, but with heavier loads the onset of shortening is progressively delayed. Finally, with sufficiently heavy loads there is no shortening at all and the contraction becomes isometric. In addition, with heavier loads the duration of shortening will be less and the maximum speed of shortening will be slower. In a sense, the measurement of tension during an isometric contraction gives a more direct view of the contractile state of the muscle; for this reason, subsequent examples in this chapter will be of isometric contractions.

Control of Muscle Tension

Recruitment of Motor Neurons

A single muscle typically receives inputs from hundreds of motor neurons. Thus, tension in the muscle can be increased by increasing the number of these motor neurons that are firing action potentials; the tension produced by activating individual motor units sums to produce the total tension in the muscle. A simplified example is shown in Figure 10-3. The increase in the number of active motor neurons is called **recruitment** of motor neurons and is an important physiological means of controlling muscle tension. When motor neurons are recruited into action during naturally occurring motor behavior, such as locomotion or lifting loads, the order of recruitment is determined by the size of the motor unit. As the tension in a muscle is increased, motor units containing a small number of muscle fibers are the first to be recruited; larger motor units are recruited later. Thus, when there is little activity in the pool of motor neurons controlling a muscle and the tension in the muscle is low, small motor units are recruited to produce an increase in tension. This insures that the added increments of tension are small and prevents large jerky increases in tension when the tension is small. As tension increases, however, further increases in tension must be larger in order to make a significant

Figure 10-3
A simple muscle consisting of four motor units of varying size. The graph below shows isometric tension in response to simultaneous action potentials (at the arrow) in various combinations of the motor neurons.

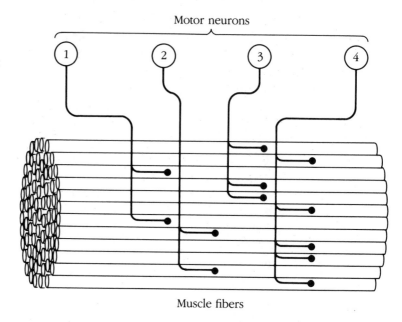

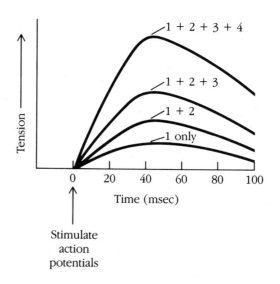

difference; thus, larger motor units are added, resulting in larger increments of tension when the background tension is already high. This behavior is referred to as the **size principle** in motor neuron recruitment. In Figure 10-3, for example, it would be expected that tension would be increased by adding the smaller motor units (1 and 2) first and the largest unit (4) last.

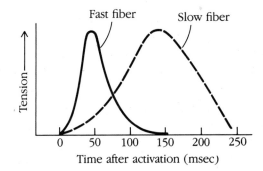

Figure 10-4
Comparison of speed of development of isometric tension in fast and slow muscle fibers.

Fast and Slow Muscle Fibers

The time delay between the occurrence of the muscle fiber action potential and the peak of the resulting tension is not constant across all muscle fibers. The delay to peak tension can be as little as 10 or as many as 200 msec. In general, muscle fibers can be grouped into two classes—fast and slow—on the basis of this speed. Samples of isometric contractions in fast and slow fibers are shown in Figure 10-4. Both slow and fast fibers are found together in most muscles, but slow fibers predominate in muscles that must maintain steady contraction, such as those involved in keeping us standing upright. Fast muscle fibers are more common in muscles that require rapid contraction, such as those involved in jumping and running. The fastest muscle fibers are those of muscles that move the eyes in rapid jumps, like those your eyes are making as you scan the words on this page.

Temporal Summation of Contractions Within a Single Motor Unit

When motor neurons are activated during naturally occurring movement, they do not typically fire just a single action potential, as has been the case in all our examples so far. Rather, action potentials tend to occur in bursts of several or as steady discharges at a relatively constant frequency. It is not uncommon for action potentials within a burst to be separated by only 10 msec or less. Under normal conditions, each of these many action potentials in a motor neuron will produce a corresponding action potential in each of the motor unit's muscle fibers. Because the tension resulting from a single action potential typically lasts for many tens or hundreds of milliseconds, there is considerable opportunity for summation of the effects of succeeding muscle action potentials in a series, as

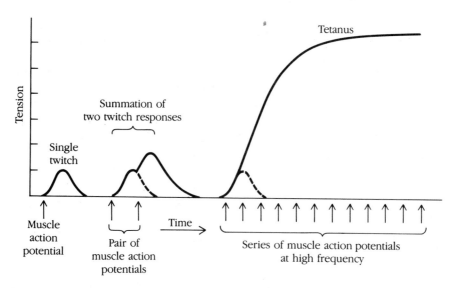

Figure 10-5

Isometric tension in response to a series of action potentials in a muscle. The dashed lines show the expected response if only the first action potential of a series occurred.

illustrated in Figure 10-5. Such temporal summation of individual twitches is a major way the nervous system controls tension in skeletal muscles.

The amount of summation within a burst of muscle action potentials depends on the frequency of action potentials: the higher the frequency, the greater the resulting summed tension. However, as shown in Figure 10-5, when the frequency is sufficiently high the individual tension responses of the muscle fuse together into a plateau of tension. Further increase in frequency beyond this point does not increase tension: the muscle has reached its maximum response and cannot develop further tension. This plateau state is called **tetanus**. As expected from the examples shown in Figure 10-4, the frequency of stimulation required to produce tetanus varies considerably depending on whether slow or fast fibers are involved. For fast fibers, a frequency of more than 100 action potentials per second may be required, while for slow fibers a frequency of 20 per second may suffice.

Asynchronous Activation of Motor Units During Maintained Contraction

As anyone who has done prolonged physical labor or exercised vigorously can attest, muscle contractions cannot be maintained indefinitely; muscles fatigue and must be rested. Thus, the state of tetanus in Figure 10-5 could not be maintained in a single motor unit for very long without allowing the muscle fibers in the motor

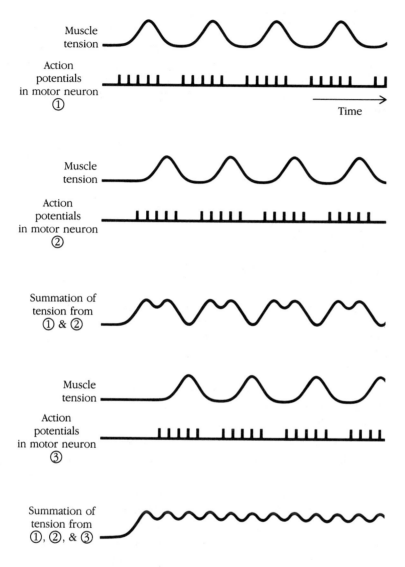

Figure 10-6
Summation of muscle tension during asynchronous activation of three motor units in a muscle.

unit to relax. However, some muscles—such as those involved in maintaining body posture—are required to contract for prolonged periods. What mechanism helps prevent muscle fatigue during such prolonged contractions? During maintained tension in a muscle, all the motor neurons to the muscle are not active at the same time. The activity of the motor units occurs in bursts separated by quiet periods, and the activity of different motor units is staggered in time. An example of this kind of asynchronous activity during steady contraction is shown in Figure 10-6. Notice that the summation of the tensions produced by the activity of only three motor units, each active only half the time, can produce a reasonably smooth, steady tension. With hundreds of motor units available in many muscles, a much smoother and larger steady tension could be maintained with less effort on the part of any one motor unit. Thus, asynchronous activation of motor neurons to a muscle allows a prolonged contraction with reduced fatigue of individual motor units in the muscle.

Summary

The basic unit of contraction of a skeletal muscle is the contraction of the group of muscle fibers making up a single motor unit, which consists of a single motor neuron and all the muscle fibers receiving synaptic connections from that neuron. Whenever the motor neuron fires an action potential, all the muscle fibers in that motor unit twitch together. The magnitude of the contraction generated by activation of a motor unit depends on the number of muscle fibers in that motor unit. The number of fibers in a unit, and hence the magnitude of the tension produced by activating the unit, varies considerably among the set of motor neurons innervating a particular muscle.

The type of contraction produced by activation of a whole muscle depends on the load against which the muscle is contracting. If the load is too great for the muscle to move, the length of the muscle does not change during the contraction, which is then called an isometric contraction. If the tension is sufficient to overcome the weight of the load, the contraction will be accompanied by a shortening of the muscle. During the shortening, the tension in the muscle remains constant and equal to the weight of the load. Such a contraction is called isotonic. The overall tension developed by a muscle depends on how many motor units are activated and on the frequency of action potentials within a motor unit. Increasing muscle tension by increasing the number of active motor neurons is called motor neuron recruitment. When the frequency of action potentials within a motor unit is increased, the resulting muscle tension increases until a steady plateau state, called tetanus, is reached. Normally, during a maintained contraction all the motor neurons of a muscle are not active simultaneously; rather, the activity of individual motor neurons is restricted to periodic bursts that occur asynchronously among the pool of motor neurons controlling a muscle. This helps reduce fatigue in the muscle by allowing individual motor units to rest periodically during a maintained contraction.

11

Electrical Properties of Cardiac Muscle

This chapter will be concerned with the properties of a special class of striated muscle cells: those that make up the muscle of the heart. These muscle cells contain a contractile apparatus like that of other striated muscle, being made up of bundles of myofilaments with a microscopic structure like that discussed in Chapter 9. Unlike other striated muscles in the body, the heart muscle is specialized to produce a rhythmic and coordinated contraction in order to drive the blood efficiently through the blood vessels. To set the stage for a discussion of the specialized cellular properties of heart muscle, we will start with a description of the cycle of cardiac contraction. Then we will turn to a description of the cellular mechanisms underlying the generation of the efficient pumping action of the heart.

The Pattern of Cardiac Contraction

As we are all aware, the heart is responsible for moving blood through the vessels of the circulatory system. It has a number of tasks to accomplish in order to carry out its role in providing oxygen to the cells of the body. It must receive the oxygen-poor blood returning from the body tissues via the venous circulation and send that blood to the lungs for oxygenation. It must also receive the oxygenated blood from the lungs and send it out through the arterial circulation to the rest of the body. Carrying out these tasks requires precise timing of the contractions of the various heart chambers; otherwise, the flow of oxygenated blood will not occur efficiently or will cease altogether—with disastrous consequences. We will consider first the normal timing sequence of the heart contraction, to see how the rhythmic contractions are coordinated to pump the blood.

A schematic diagram of the flow of blood through a human heart during a single contraction cycle is shown in Figure 11-1. Humans, like all other mammals, have a four-chambered heart, consisting of the left and right atria and the left and right ventricles. The two atria can be thought of as the receiving chambers, or "priming" pumps, of the heart, while the two ventricles are the "power" pumps of the circulatory system. The right atrium receives the blood returning from the body through the veins, and the left atrium receives the freshly oxygenated blood from the lungs. During the phase of the heartbeat when the atria are filling with blood, the valves connecting the atria with the ventricles are closed, preventing flow of blood into the ventricles. When the atria have filled with blood, they contract and the increase in pressure opens the valves leading to the ventricles and drives the collected blood into the ventricles. At this point, the muscle of the ventricles is relaxed, and the valves connecting the ventricles to the vessels leaving the heart are closed. When the ventricles have filled with blood, they contract, opening these valves and delivering the power stroke to drive the blood out to the lungs and to the rest of the body, as shown in Figure 11-1B. Thus, during a normal heartbeat the two atria contract together, followed after a delay by the simultaneous contraction of the two ventricles.

Coordination of Contraction Across Cardiac Muscle Fibers

In order for the contraction of a heart chamber to be able to propel the expulsion of fluid, all the individual muscle fibers making up the walls of that chamber must contract together. It is this unified contraction that constricts the cavity of the chamber and drives out the blood into the blood vessels of the circulation. In skeletal muscles, an action potential in one muscle fiber is confined to that fiber and does not influence neighboring fibers; therefore, contraction is restricted to the particular fiber undergoing an action potential. In cardiac muscle, however, the situation is quite different. When an action potential is generated in a cardiac muscle fiber, it causes action potentials in the neighboring fibers, which in turn set up action potentials in their neighbors, and so on. Thus, the

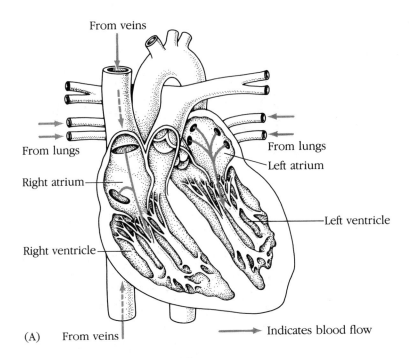

From veins

From lungs

Right atrium

Right ventricle

From lungs

Left atrium

Left ventricle

(A) From veins

→ Indicates blood flow

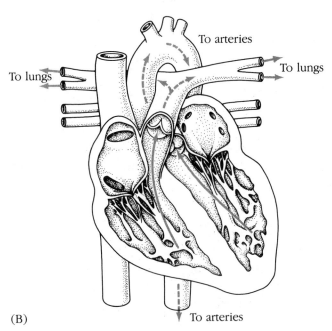

To arteries

To lungs

To lungs

(B) To arteries

Figure 11-1
Schematic drawings of the state of the heart valves and the direction of blood flow during two stages in a single heartbeat. In (A), the atria are contracting and the ventricles are filling with blood. In (B), the valves between the atria and ventricles are closed and the ventricles are contracting, forcing the blood out to the lungs (right ventricle) and to the arteries leading to the rest of the body (left ventricle).

Figure 11-2
Electrical current can flow from
one cardiac muscle cell to
another through specialized
membrane junctions located in a
region of contact called the
intercalated disc.

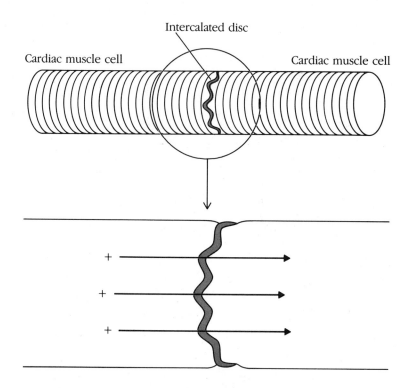

excitation spreads rapidly out through all the muscle fibers of the
chamber. This insures that all the fibers contract together.

What characteristic of cardiac muscle fibers allows the action
potential to spread from one fiber to another? This can be seen by
looking at the microscopic structure of the cells of cardiac muscle,
as shown schematically in Figure 11-2. At the ends of each cardiac
cell, the plasma membranes of neighboring cells come into close
contact at specialized structures called **intercalated discs**. The
contact at this point is sufficiently close that electrical current flowing inside one fiber can cross directly into the interior of the next
fiber; in electrical terms, it is as though the neighboring cells form
one larger cell. Recall from Chapter 5 that an electrical current
flowing along the interior of a fiber has at each point two paths to
choose from: across the plasma membrane or continuing along the
interior of the fiber. The amount of current taking each path at a
particular point depends on the relative resistances of the two
paths; the higher the resistance, the smaller the amount of current
taking that path. Normally, at the point where one cell ends and the
next begins, there is little opportunity for current to flow from one
cell to the other because the current would have to flow out across
one cell membrane and in across the other in order to do so; this is

a high resistance path because current must cross two membranes. At the specialized structure of the intercalated disc, however, the resistance to current flow across the two membranes is low, so that the path to the interior of the neighboring cell is favored. This means that depolarizing current injected into one cell during the occurrence of an action potential can spread directly into neighboring cells, setting up an action potential in those cells as well. The low resistance path from one cell to another is through membrane structures called **gap junctions**. These structures appear to consist of arrays of small pores directly connecting the interiors of the joined cells, so that small molecules like ions can pass directly from one cell to another.

When electrical current can pass from one cell to another, as in cardiac muscle, those cells are said to be **electrically coupled**. An electrophysiological experiment to demonstrate this behavior is illustrated in Figure 11-3. When current is injected into a cell, no response occurs in a neighboring cell if the cells are not electrically coupled. If the two cells are coupled via gap junctions, a response to the injected current occurs in both cells because the ions carrying the current inside the cell can pass directly through the gap junction. If the depolarization is large enough, an action potential will be triggered in both cells at the same time.

Generation of Rhythmic Contractions

The electrical coupling among cardiac muscle fibers can explain how contraction occurs synchronously in all the fibers of a chamber. We will now consider the control mechanisms responsible for the repetitive contractions that characterize the beating of the heart. If a heart is removed from the body and placed in an appropriate artificial environment, it will continue to contract repetitively even though it is isolated from the nervous system and the rest of the body. In contrast, a skeletal muscle isolated under similar conditions will not contract unless its nerve is activated. The rhythmic activity of the heart muscle is an inherent property of the individual muscle fibers making up the heart, and this constitutes another important difference between cardiac muscle fibers and skeletal muscle fibers. This difference can be demonstrated dramatically in experiments in which muscle tissue is dissociated into individual cells, which are placed in a dish isolated from each other and from

Figure 11-3
Results of an experiment in which the membrane potentials of two cells are measured simultaneously with intracellular microelectrodes. (A) A depolarizing current is injected into cell 1. (B) If the cells are not electrically coupled, the depolarization occurs only in the cell in which the current was injected. (C) If the cells are electrically coupled via gap junctions, a depolarization occurs in cell 2 as well as in cell 1.

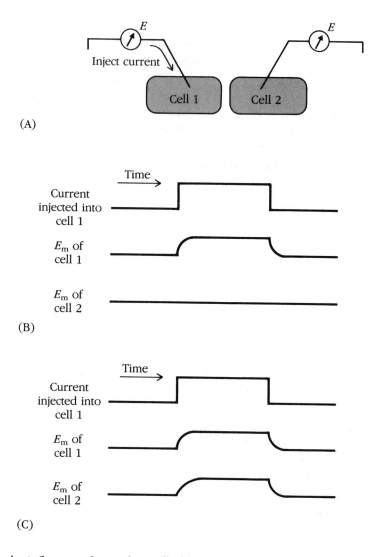

the influence of any other cells, like nerve cells. Under these conditions, muscle cells from skeletal muscles are quiescent; they do not contract in the absence of their neural input. Cells from cardiac muscle, however, continue to contract rhythmically even in isolation. Thus, rhythmic contractions of heart muscle are due to built-in properties of the cardiac muscle cells. Before we can examine the membrane mechanism underlying this **autorhythmicity**, it will be necessary to look first at the action potential of cardiac muscle cells. In keeping with the different behavior of cardiac cells, this action potential has some different characteristics from the action potentials of neurons or skeletal muscle cells.

The Cardiac Action Potential

In Chapters 5 and 6, we discussed the ionic mechanisms underlying the action potential of nerve membrane. The action potential of skeletal muscle fibers is fundamentally the same as that of neurons. The cardiac action potential, however, is different from these other action potentials in several important ways. Figure 11-4 compares the characteristics of action potentials of skeletal and cardiac muscle cells. One striking difference is the difference in time-scale: cardiac action potentials can last several hundred milliseconds, while skeletal muscle action potentials are typically over in 1 to 2 msec. The underlying changes in membrane permeability shown in Figure 11-4 demonstrate that the long-lasting plateau of the cardiac action potential is due to two factors: a prolonged increase in the calcium permeability of the plasma membrane and a simultaneous reduction in the potassium permeability. In addition, the sodium permeability does not completely shut off, as it does in nerve or skeletal muscle action potentials. The combination of increased calcium permeability and reduced potassium permeability serves to maintain the depolarization during the plateau phase of the cardiac action potential.

The increase in calcium permeability is due to the presence of voltage-sensitive calcium channels in the plasma membrane of the cardiac muscle fiber. Like the voltage-sensitive sodium channels of nerve membrane (see Chapter 5), these calcium channels open upon depolarization, although they do so slowly. Unlike the sodium channels, the calcium channels do not rapidly close again with maintained depolarization. From the Nernst equation, it can be calculated that the calcium equilibrium potential is strongly positive (about +100 to +150 mV); therefore, an increase in calcium permeability will tend to keep the cardiac muscle cell depolarized. This accounts in part for the long plateau phase of the cardiac action potential.

The plateau of the cardiac action potential is also associated with a reduction in the potassium permeability. This is due to an unusual type of potassium channel that is open as long as the membrane potential is near its normal resting level and *closes* upon depolarization. This is the reverse of the behavior of the gated potassium channel we are familiar with from our discussion of nerve action potential. The reduction in potassium permeability caused by the closing of this channel tends to depolarize the cardiac muscle fiber. In fact, the depolarization caused by this unusual

Figure 11-4
The sequence of permeability changes underlying the action potentials of (A) skeletal muscle fibers and (B) cardiac muscle fibers. Note the greatly different time-scales.

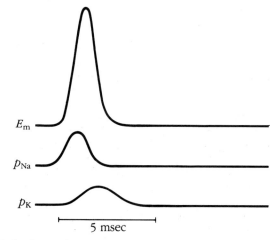

E_m

p_{Na}

p_K

⊢————— 5 msec —————⊣

(A)   Skeletal muscle

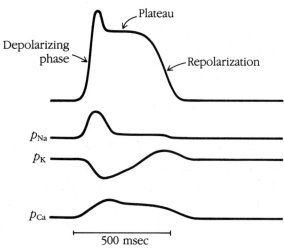

Depolarizing phase ↘   Plateau

Repolarization

p_{Na}

p_K

p_{Ca}

⊢————— 500 msec —————⊣

(B)   Cardiac muscle

potassium channel is **regenerative**, just like the explosive depolarization caused by the sodium channel of nerve action potentials. That is, depolarization shuts some of the potassium channels, which reduces potassium permeability and causes more depolarization, which shuts more potassium channels, and so on. Thus, it turns out that the plateau phase of the cardiac action potential can still occur even if the voltage-sensitive calcium channels mentioned above are blocked by drugs. This shows that the closing of the potassium channels is by itself capable of producing the prolonged plateau of the cardiac action potential. Normally, however, both the opening of the calcium channels and the closing of the potassium channels contribute to the plateau.

The initial rising phase of the cardiac action potential is due to a sodium channel very much like that of nerve membrane. This channel drives the rapid initial depolarization and is responsible for the brief initial spike of the cardiac action potential before the plateau phase sets in. Like the sodium channel of neuronal membrane, this channel rapidly closes (inactivates) with maintained depolarization. However, unlike the nerve sodium channel, this inactivation is not total; there is a small, maintained increase in sodium permeability during the plateau.

What is responsible for terminating the cardiac action potential? First, the calcium permeability of the plasma membrane slowly declines during the maintained depolarization. This decline might be a consequence of the gradual build up of internal calcium concentration as calcium ions continue to enter the muscle fiber through the open calcium channels. Internal calcium ions are thought to have some direct action on the calcium channels, causing them to close. In addition, the potassium permeability of the plasma membrane increases, as in the nerve and skeletal muscle action potentials. This increase in potassium permeability tends to drive the membrane potential of the cardiac fiber toward the potassium equilibrium potential and thus to repolarize the fiber. There is evidence that part of this increase in potassium permeability is due to voltage-sensitive potassium channels that open in response to the depolarization during the action potential (like the n gates of the nerve membrane). However, part of the increased potassium permeability appears to be caused by another type of potassium channel that is opened not by depolarization but by an increase in internal calcium concentration. Thus, the rise in internal calcium concentration during the prolonged action potential helps shut off the action potential not only by reducing calcium permeability but also by increasing potassium permeability. The different kinds of membrane ionic channels underlying the cardiac action potential are summarized in Table 11-1.

One functional implication of the prolonged cardiac action potential is that the duration of the contraction in cardiac muscle is controlled by the duration of the action potential. The action potential and contraction of cardiac muscle fibers are compared with those of skeletal muscle fibers in Figure 11-5. In skeletal muscle, the action potential acts only as a trigger for the contractile events;

Table 11-1   Summary of ionic channels of cardiac muscle membrane

Principal ion	Response to depolarization	Speed of response	Inactivation	Function
Na	Opens	Fast	Fast, but incomplete	Initial depolarization
K	Closes	Fast	None	Maintains plateau
K	Opens	Slow	Little	Repolarization
K	Opens due to Ca influx	Slow	Closes as internal Ca falls	Repolarization
Ca	Opens	Slow	Slow	Maintains plateau and prolongs contraction

Figure 11-5
(A) In a skeletal muscle fiber, the action potential is much briefer than the resulting contraction. Thus, the action potential acts only as a trigger for the contraction, which proceeds independently of the duration of the action potential. (B) In a cardiac muscle fiber, the duration of the contraction is closely related to the duration of the action potential because of the maintained calcium influx during the plateau of the action potential. Thus, characteristics of the action potential can influence the duration and strength of the cardiac contraction.

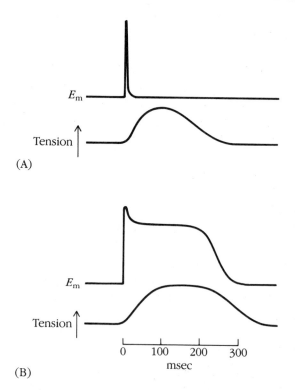

the duration of the contraction is controlled by the timing of the release and reuptake of calcium by the sarcoplasmic reticulum. In cardiac muscle fibers, however, only the initial part of the contraction is controlled by sarcoplasmic reticulum calcium; the contraction is maintained by the influx of calcium ions across the

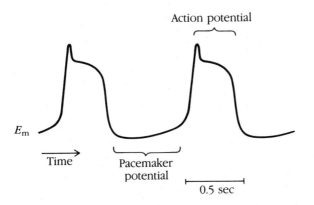

Action potential

E_m

Time

Pacemaker
potential

0.5 sec

Figure 11-6
Recording of membrane potential
during repetitive, spontaneous
beating in a single cardiac muscle
fiber. The repolarization at the
end of one action potential is
followed by a slow, spontaneous
depolarization called the
pacemaker potential. When this
depolarization reaches threshold,
a new action potential is
triggered.

plasma membrane during the plateau phase of the cardiac action
potential. For this reason, the duration of the contraction in the
heart can be altered by changing the duration of the action potential in the cardiac muscle fibers. This provides an important mechanism by which the pumping action of the heart can be modulated.

The Pacemaker Potential

Although the ionic mechanism of the cardiac action potential differs
in important ways from that of other action potentials, nothing in
the scheme presented so far would account for the endogenous
beating of isolated heart cells discussed earlier. If we recorded the
electrical membrane potential of a spontaneously beating isolated
heart cell, we would see a series of spontaneous action potentials,
as shown in Figure 11-6. After each action potential, the potential
falls to its normal negative resting value, then begins to depolarize
slowly. This slow depolarization is called a **pacemaker potential**,
and it is due to spontaneous changes in the membrane ionic permeability. Voltage clamp experiments on single isolated muscle
fibers from the ventricles suggest that the pacemaker potential is
due to a slow decline in the potassium permeability coupled with a
slow increase in sodium permeability. When the depolarization
reaches threshold, it triggers an action potential in the fiber. Unfortunately, the molecular mechanisms underlying the spontaneous changes in sodium and potassium permeability are not yet
understood.

The rate of spontaneous action potentials in isolated heart cells
varies from one cell to another; some cells beat rapidly and others
slowly. In the intact heart, however, the electrical coupling among
the fibers guarantees that all the fibers will contract together, with

Figure 11-7

Diagram of the spread of action potentials across the heart during a single heartbeat. The excitation originates in the sinoatrial (SA) node of the right atrium and spreads throughout the atria via electrical coupling among the atrial muscle fibers. The spread is facilitated by special conducting fibers in Bachmann's bundle. The fibers of the atria are not electrically connected to those of the ventricles. The action potential spreads to the ventricles via the atrioventricular (AV) node, which introduces a delay between the atrial and ventricular action potentials. When the wave of action potentials leaves the AV node, its spread throughout the ventricles is aided by the rapidly conducting Purkinje fibers of the bundle of His.

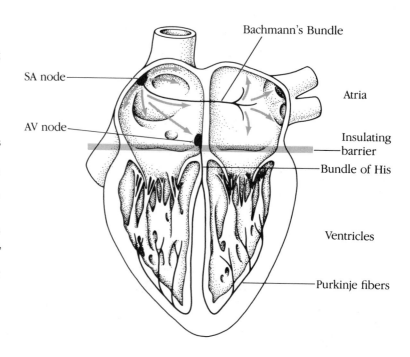

the rate being governed by the fibers with the fastest pacemaker activity. In the normal functioning of the heart, the overall rate of beating is controlled by a special set of pacemaker cells, called the **sinoatrial (SA) node**, which is located in the upper part of the right atrium. This node is indicated in the diagram of the heart in Figure 11-7. In the normal resting heart, the cells of the SA node generate spontaneous action potentials at a rate of about 70 per minute. These action potentials spread through the electrical connections among fibers throughout the two atria, generating the simultaneous contraction of the atria as discussed in the first section of this chapter. The spread of excitation from the right to the left atrium is facilitated by a bundle of muscle fibers, called **Bachmann's bundle**, that are specialized for more rapid conduction of excitation. This helps insure that the two atria contract together. The atrial action potentials do not spread directly to the fibers making up the two ventricles, however. This is a good thing, because we know that the contraction of the ventricles must be delayed to allow the relaxed ventricles to fill with blood pumped out by the atrial contraction. In terms of electrical conduction, the heart behaves as two isolated units, as shown in Figure 11-7: the two atria are one unit and the two ventricles are another. The electrical connection between these two units is made via another specialized group of muscle fibers called the **atrioventricular**

(AV) node. Excitation in the atria must travel through the AV node to reach the ventricles. The fibers of the AV node are small in diameter compared with other cardiac fibers; as discussed in Chapter 5, the speed of action potential conduction is slow in small diameter fibers. Therefore, conduction through the AV node introduces a time delay sufficient to retard the contraction of the ventricles relative to the contraction of the atria. Excitation leaving the AV node does not travel directly through the muscle fibers of the ventricles. Instead, it travels along specialized muscle fibers that are designed for rapid conduction of action potentials. These fibers are called **Purkinje fibers**, and they form a fast-conducting pathway through the ventricles called the **bundle of His**. The Purkinje fibers carry the excitation rapidly to the apex of the heart, where it then spreads out through the mass of ventricular muscle fibers to produce the contraction of the ventricles.

Neural Control of the Heart

Each muscle fiber of a skeletal muscle receives a direct synaptic input from a particular motor neuron; without this synaptic input, a skeletal fiber does not contract unless stimulated directly by artificial means. On the other hand, we have seen that cardiac muscle fibers generate spontaneous contractions that are coordinated into a functional heartbeat by the electrical conduction mechanisms inherent in the heart itself. This does not mean, however, that the activity of the heart is not influenced by inputs from the nervous system. The heart receives two opposing neural inputs that affect the heart rate. One input comes from the cells of the parasympathetic nervous system, whose synaptic terminals in the heart release the neurotransmitter acetylcholine. The effect of acetylcholine is to slow the rate of depolarization during the pacemaker potential of the SA node. This has the effect of increasing the interval between successive action potentials and thus slowing the rate at which this master pacemaker region drives the heartbeat. Acetylcholine appears to act by increasing the potassium permeability of the muscle fiber membrane. This tends to keep the membrane potential closer to the potassium equilibrium potential and thus retard the growth of the pacemaker potential toward threshold for triggering an action potential.

The second neural input to the heart comes from cells of the sympathetic nervous system, whose synaptic terminals release the neurotransmitter norepinephrine. Activation of this input speeds

Table 11-2 Comparison of some properties of skeletal and cardiac muscle fibers

Property	Skeletal muscle	Cardiac muscle
Striated	Yes	Yes
Electrically coupled	No	Yes
Spontaneously contract in absence of nerve input	No	Yes
Duration of contraction controlled by duration of action potential	No	Yes
Action potential is similar to that of neurons	Yes	No
Calcium ions make an important contribution to the action potential	No	Yes
Effect of neural input	excite	excite or inhibit

the heart rate. Again, this effect is mediated via the pacemaker potential, which depolarizes more rapidly after activation of the sympathetic input. The mechanism of this effect is not yet understood in detail, although it seems to involve a simultaneous decrease in potassium permeability and increase in calcium permeability. Both of these factors would tend to accentuate the depolarizing trend during the pacemaker potential.

Summary

The muscle fibers making up the heart are specialized in a number of ways to carry out their function of efficiently pumping blood through the vessels of the circulatory system. These specializations lead to a number of differences between cardiac muscle fibers and skeletal muscle fibers, which are summarized in Table 11-2. In addition, the heart as an organ contains specific structures whose function is to coordinate the pumping activity. These structures include the SA node, the AV node, and the Purkinje fibers. The SA node is the master pacemaker region of the heart, which controls the heart rate during normal physiological functioning of the heart. The AV node provides a path for electrical conduction between the atria and the ventricles and is responsible for the delay between atrial and ventricular contractions. The Purkinje fibers provide a rapidly conducting pathway for distributing excitation throughout the ventricles during the power stroke of a single heartbeat.

Derivation of the Nernst Equation

The Nernst equation is used extensively in the discussion of resting membrane potential and action potentials in this book. The derivation presented here is necessarily mathematical and requires some knowledge of differential and integral calculus to understand thoroughly. However, I have tried to explain the meaning of each step in words; hopefully, this will allow those without the necessary background to follow the logic qualitatively.

This derivation of the Nernst equation uses equations for the movement of ions down concentration and electrical gradients to arrive at a quantitative description of the equilibrium condition. The starting point is the realization that at equilibrium there will be no net movement of the ion across the membrane. In the presence of both concentrational and electrical gradients, this means that the rate of movement of the ion down the concentration gradient is equal and opposite to the rate of movement of the ion down the electrical gradient. For a charged substance (an ion), movement across the membrane constitutes a transmembrane electrical current, I. Thus, at equilibrium

$$I_C = -I_E \tag{A-1}$$

or

$$I_C + I_E = 0 \tag{A-2}$$

where I_C and I_E are the currents due to the concentrational and electrical gradients, respectively.

Concentrational Flux

Consider first the current due to the concentration gradient, which will be given by

$$I_C = A\Phi_C ZF \qquad (A\text{-}3)$$

In words, equation (A-3) states that the current through the membrane of area A will be equal to the flux, Φ_C, of the ion down the concentration gradient (number of ions per second per unit area of membrane) multiplied by Z (the valence of the ion) and F (Faraday's constant; 96,500 coulombs per mole of univalent ion). The factor ZF translates the flux of ions into flux of charge and hence into an electrical current. The flux Φ_C for a given ion (call the ion Y, for example) will depend on the concentration gradient of Y across the membrane (that is, $[Y]_{in} - [Y]_{out}$) and on the membrane permeability to Y, p_Y. Quantitatively, this relation is given by

$$\Phi_C = p_Y([Y]_{in} - [Y]_{out}) \qquad (A\text{-}4)$$

Note that p_Y has units of velocity (cm/sec), in order for Φ_C to have units of molecules/sec/cm² (remember that $[Y]$ has units of molecules/cm³). The permeability coefficient, p_Y, is in turn given by

$$p_Y = D_Y/a \qquad (A\text{-}5)$$

where D_Y is the diffusion constant for Y within the membrane and a is the thickness of the membrane. D_Y can be expanded to yield

$$D_Y = uRT \qquad (A\text{-}6)$$

where u is the mobility of the ion within the membrane and RT (the gas constant times the absolute temperature) is the thermal energy available to drive ion movement. Substituting equation (A-6) in (A-5) and the result in (A-4) yields

$$\Phi_C a = uRT([Y]_{in} - [Y]_{out}) \qquad (A\text{-}7)$$

Equation (A-7) gives us the flux through a membrane of thickness a, but we would like a more general expression that gives us the flux through any arbitrary plane in the presence of a concentration gradient. To arrive at this expression, consider the sit-

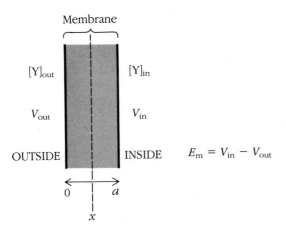

Figure A-1
Segment of membrane separating
two compartments.

uation diagrammed in Figure A-1, which shows a segment of membrane separating two compartments. The dimension perpendicular to the membrane is called x, and the membrane extends from 0 to a (thickness $= a$). In this situation, equation (A-7) can be expressed in the form of an integral equation:

$$\Phi_C\left(\int_0^a dx\right) = uRT\left(\int_0^a dC\right) \qquad \text{(A-8)}$$

Here, C stands for the concentration of the ion; therefore, in reference to Figure A-1, C_a is $[Y]_{in}$ and C_0 is $[Y]_{out}$. Differentiating both sides of equation (A-8) yields

$$\Phi_C \, dx = uRT \, dC \qquad \text{(A-9)}$$

which can be arranged to give the more general form of equation (A-7) that we desire:

$$\Phi_C = uRT \left(\frac{dC}{dx}\right) \qquad \text{(A-10)}$$

Equation (A-10) can be substituted into equation (A-3) to give us the ionic current due to the concentration gradient.

Current Due to Electrical Gradient

Return now to the current driven by the electrical gradient, which can be expressed

$$I_E = A\Phi_E ZF \tag{A-11}$$

The flux, Φ_E, of a charged particle through a plane at position x in the presence of a voltage gradient dV/dx will be

$$\Phi_E = uZFC\left(\frac{dV}{dx}\right) \tag{A-12}$$

Again, u is the mobility of the ion, and C is the concentration of the ion at position x. The factor ZFC is then the concentration of charge at the location of the plane; this is necessary because the voltage gradient dV/dx acts on charge and ZFC gives the "concentration" of charge at position $= x$. Equation (A-12) is analogous to equation (A-10), except it is the voltage gradient rather than the concentration gradient that is of interest.

Total Current at Equilibrium

Equations (A-12), (A-11), (A-10) and (A-3) can be combined into the form of equation (A-2) to give

$$uAZF\left(RT\frac{dC}{dx} + ZFC\frac{dV}{dx}\right) = 0 \tag{A-13}$$

This requires that

$$RT\left(\frac{dC}{dx}\right) = -ZFC\left(\frac{dV}{dx}\right) \tag{A-14}$$

Equation (A-14) can be rearranged to give a differential equation that can be solved for the equilibrium voltage gradient:

$$\left(-\frac{RT}{ZF}\right)\left(\frac{dC}{C}\right) = dV \tag{A-15}$$

This can be solved for V by integrating across the membrane. Using the nomenclature of Figure A-1, the integrals are

$$-\frac{RT}{ZF}\int_{[Y]_{out}}^{[Y]_{in}} \frac{dC}{C} = \int_{V_{out}}^{V_{in}} dV \tag{A-16}$$

The solution to these definite integrals is

$$-\frac{RT}{ZF}(\ln[Y]_{in} - \ln[Y]_{out}) = V_{in} - V_{out} \qquad \text{(A-17)}$$

or

$$\frac{RT}{ZF}\ln\left(\frac{[Y]_{out}}{[Y]_{in}}\right) = V_{in} - V_{out} = E_M \qquad \text{(A-18)}$$

Equation (A-18) is the Nernst equation.

Derivation of the Goldman Equation

The Goldman equation, or constant-field equation, is important to an understanding of the factors that govern the steady-state membrane potential. As discussed in Chapter 4, the Goldman equation describes the nonequilibrium membrane potential reached when two or more ions with unequal equilibrium potentials are free to move across the membrane. The basic strategy in this derivation is to use the flux equations derived in Appendix A to solve separately for the ionic current carried by each permeant ion and then to set the sum of all ionic currents equal to zero. The derivation is somewhat more complex than that of the Nernst equation in Appendix A, and it requires some knowledge of differential and integral calculus to follow in detail. Nevertheless, it should be possible for those without the necessary mathematics to follow the logic of the steps and thus to gain some insight into the physical mechanisms described by the equation.

When several ions are moving across the membrane simultaneously, a steady value of membrane potential will be reached when the sum of the ionic currents carried by the individual ions is zero; that is, for permeant ions A, B, and C

$$I_A + I_B + I_C = 0 \qquad \text{(B-1)}$$

The first step in arriving at a value of membrane potential that satisfies this condition is to solve for the net ionic flux, Φ, for each ion separately. The total flux for a particular ion will be the sum of the flux due to the concentration gradient and the flux due to the electrical gradient:

$$\Phi_T = \Phi_C + \Phi_V \qquad \text{(B-2)}$$

The expressions for Φ_C and Φ_V are given by equations (A-10) and (A-12) in Appendix A. Thus, equation (B-2) becomes

$$\Phi_T = uRT\,(dC/dx) + uZFC\,(dV/dx) \tag{B-3}$$

If it is assumed that the electric field across the membrane is constant (this is the constant-field assumption from which the equation draws its alternative name) and that the thickness of the membrane is a, then

$$dV/dx = V/a \tag{B-4}$$

In that case, equation (B-3) can be written

$$\frac{\Phi_T}{uRT} = \frac{dC}{dx} + \frac{ZFV}{RTa}C \tag{B-5}$$

This is a differential equation of the form

$$Q = \frac{dC}{dx} + P(x)C$$

which has a solution

$$C \exp\left(\int P(x)\,dx\right) = \int Q \exp\left(\int P(x)\,dx\right) dx + \text{constant} \tag{B-6}$$

In this instance, $Q = \Phi_T/(uRT)$ and $P(x) = (ZFV)/(RTa)$. Making these substitutions and integrating equation (B-6) across the membrane of thickness a (that is, from 0 to a) gives

$$C \exp\left(\frac{ZFV}{RTa}\right)\Big|_0^a = \frac{\Phi_T}{uRT} \int_0^a \exp\left(\frac{ZFVx}{RTa}\right) dx \tag{B-7}$$

This becomes

$$C_a \exp\left(\frac{ZFV}{RT}\right) - C_0 = \frac{\Phi_T}{uRT}\left[\exp\left(\frac{ZFVx}{RTa}\right)\bigg/\left(\frac{ZFV}{RTa}\right)\right]\Big|_0^a$$

or

$$C_a \exp\left(\frac{ZFV}{RT}\right) - C_0 = \frac{\Phi_T}{uRT}\frac{RTa}{ZFV}\left[\exp\left(\frac{ZFVa}{RTa}\right) - \exp\left(\frac{ZFV \cdot 0}{RTa}\right)\right]$$

Rearranging and combining terms yields

$$C_a \exp\left(\frac{ZFV}{RT}\right) - C_0 = \frac{\Phi_T a}{uZFV}\left[\exp\left(\frac{ZFV}{RT}\right) - 1\right]$$

This can be solved for Φ_T to yield

$$\Phi_T = \frac{uZFV}{a}\left[\frac{C_a \exp(ZFV/RT) - C_0}{\exp(ZFV/RT) - 1}\right] \qquad (B\text{-}8)$$

Now, C_a and C_0 are the concentrations of the ion just within the membrane. These concentrations are related to the concentrations in the fluids inside and outside the cell by $C_a = \beta C_{in}$ and $C_0 = \beta C_{out}$, where β is the oil–water partition coefficient for the ion in question. Substituting these in equation (B-8) gives

$$\Phi_T = \frac{\beta uZFV}{a}\left[\frac{C_{in}\exp(ZFV/RT) - C_{out}}{\exp(ZFV/RT) - 1}\right] \qquad (B\text{-}9)$$

The permeability constant, p_i, for a particular ion is given by $p_i = \beta uRT/a$, or $p_i/RT = \beta u/a$. Making this substitution in equation (B-9) gives

$$\Phi_T = \frac{p_i ZFV}{RT}\left[\frac{C_{in}\exp(ZFV/RT) - C_{out}}{\exp(ZFV/RT) - 1}\right] \qquad (B\text{-}10)$$

The flux, Φ_T, for an ion can be converted to a flow of electrical current, as required in equation (B-1), by multiplying by ZF (the number of coulombs per mole of ion); therefore

$$I = \frac{p_i Z^2 F^2 V}{RT}\left[\frac{C_{in}\exp(ZFV/RT) - C_{out}}{\exp(ZFV/RT) - 1}\right] \qquad (B\text{-}11)$$

This is the expression we need for each ion in equation (B-1). For instance, if the three permeant ions are Na, K, and Cl with permeabilities p_{Na}, p_K and p_{Cl}, then equation (B-1) becomes (keeping in mind that the valence of chloride is -1)

$$\frac{F^2 V}{RT}\left[\frac{p_K([K]_{in}e^{FV/RT} - [K]_{out}) + p_{Na}([Na]_{in}e^{FV/RT} - [Na]_{out})}{\exp(FV/RT) - 1} + \frac{p_{Cl}([Cl]_{in}e^{-FV/RT} - [Cl]_{out})}{\exp(-FV/RT) - 1}\right] = 0$$

Multiplying through by $-\exp(FV/RT)/-\exp(FV/RT)$ and rearranging yields

$$\frac{F^2V}{RT(\exp(FV/RT) - 1)}[(p_K[K]_{in} + p_{Na}[Na]_{in} + p_{Cl}[Cl]_{out})e^{FV/RT} - (p_K[K]_{out} + p_{Na}[Na]_{out} + p_{Cl}[Cl]_{in})] = 0$$

This requires that

$$(p_K[K]_{in} + p_{Na}[Na]_{in} + p_{Cl}[Cl]_{out})e^{FV/RT} - (p_K[K]_{out} + p_{Na}[Na]_{out} + p_{Cl}[Cl]_{in}) = 0$$

or

$$e^{FV/RT} = \frac{(p_K[K]_{out} + p_{Na}[Na]_{out} + p_{Cl}[Cl]_{in})}{(p_K[K]_{in} + p_{Na}[Na]_{in} + p_{Cl}[Cl]_{out})}$$

Taking the natural logarithm of both sides and solving for V yields the usual form of the Goldman equation

$$V = \frac{RT}{F}\ln\left(\frac{p_K[K]_{out} + p_{Na}[Na]_{out} + p_{Cl}[Cl]_{in}}{p_K[K]_{in} + p_{Na}[Na]_{in} + p_{Cl}[Cl]_{out}}\right)$$

References

This section contains suggested additional readings for those interested in obtaining more advanced information about the topics covered in this book or in reading additional introductory material. The references are divided into two sections: General References, which present more advanced information, and Specific Topics. The references given under Specific Topics represent a range of difficulty from introductory (those marked with an asterisk) to original papers in technical journals.

General References

Intermediate

Aidley, D. J. *The Physiology of Excitable Cells*. 2d ed. New York: Cambridge University Press, 1979.

Berne, R. M., and Levy, M. N. *Physiology*. St. Louis: C. V. Mosby, 1983.

Katz, B. *Nerve, Muscle and Synapse*. New York: McGraw-Hill, 1966.

Kuffler, S. W., Nicholls, J. G., and Martin, A. R. *From Neuron to Brain*. 2d ed. Sunderland, Mass.: Sinauer, 1984.

Advanced

Annual Review of Physiology, Annual Review of Neuroscience, and *Annual Review of Biophysics and Bioengineering*. Published yearly by Annual Reviews, Inc., Palo Alto, Calif. Each volume contains review articles covering recent developments in physiology and neuroscience. Intended for working scientists in the field. Typically requires good background knowledge.

Davson, H. *A Textbook of General Physiology*. 4th ed. Boston: Little, Brown, 1970.

Guyton, A. C. *Textbook of Medical Physiology*. 6th ed. Philadelphia: Saunders, 1981.

Handbook of Physiology. Volumes published periodically by the American Physiological Society. Those on neurophysiology and cardiovascular physiology contain advanced material on topics covered in this book. Articles often require advanced knowledge of biology, chemistry and mathematics.

Hille, B. *Ionic Channels of Excitable Membranes.* Sunderland, Mass.: Sinauer, 1984.

Jack, J. J. B., Noble, D., and Tsien, R. W. *Electric Current Flow in Excitable Cells.* New York: Oxford University Press, 1983. Not for the timid. Only those who are comfortable with partial differential equations, complex numbers, and Laplace transforms should trespass here.

Mountcastle, V. B. *Medical Physiology.* 14th ed. St. Louis: C. V. Mosby, 1979.

Physiological Reviews. This journal is published quarterly by the American Physiological Society. Issues often contain reviews of recent advances in nerve and muscle physiology.

Ruch, T. C., and Patton, H. D. *Physiology and Biophysics.* 20th ed. St. Louis: C. V. Mosby, 1974.

Specific Topics

Asterisks indicate easier material.

The cell and its composition

Fawcett, D. W. *The Cell.* Philadelphia: Saunders, 1966.

Fettiplace, R., and Haydon, D. A. Water permeability of lipid membranes. *Physiological Reviews* 60(1980):510.

*Folsome, C. E. *Life: Origin and Evolution.* San Francisco: Freeman, 1979. A collection of *Scientific American* reprints.

*Folsome, C. E. *The Origin of Life.* San Francisco: Freeman, 1979.

*Fox, C. F. The structure of cell membranes. *Scientific American*, Feb. 1972. Also available as Offprint 1241.

Gilles, R. *Mechanisms of Osmoregulation: Maintenance of Cell Volume.* New York: Wiley, 1979.

*Lodish, H. F., and Rothman, J. E. The assembly of cell membranes. *Scientific American*, Jan. 1979. Also available as Offprint 1415.

MacKnight, A. D. C., and Leaf, A. Regulation of cellular volume. *Physiological Reviews* 57(1977):510.

Singer, S. J. 1974. The molecular organization of membranes. *Annual Review of Physiology* 43(1974):805.

Singer, S. J., and Nicolson, G. L. The fluid mosaic model of the structure of cell membranes. *Science* 175(1972):720.

Stein, W. D. *The Movement of Molecules Across Cell Membranes.* New York: Academic Press, 1967.

Resting membrane potential

Hodgkin, A. L., and Horowicz, P. The influence of potassium and chloride ions on the membrane potential of single muscle fibers. *Journal of Physiology* 148(1959):127.

Hodgkin, A. L., and Katz, B. The effects of sodium ions on the electrical activity of the giant axon of the squid. *Journal of Physiology* 108(1949):37.

Katz, B. *Nerve, Muscle and Synapse*. New York: McGraw-Hill, 1966.

Kerkut, G. A., and York, B. *The Electrogenic Sodium Pump*. Bristol, U.K.: Scientechnica, 1971.

*Keynes, R. D. Ion channels in the nerve cell membrane. *Scientific American*, March 1979. Also available as Offprint 1423.

*Ochs, S. *Elements of Neurophysiology*. New York: Wiley, 1965.

Ruch, T. C., Patton, H. D., Woodbury, W., and Towe, A. L. *Neurophysiology*. Philadelphia: Saunders, 1965.

Action potential

*Baker, P. F. The nerve axon. *Scientific American*, March 1966. Also available as Offprint 1038.

Hodgkin, A. L. *The Conduction of the Nervous Impulse*. Springfield, Ill.: Thomas, 1963.

Hodgkin, A. L., and Huxley, A. F. Quantitative description of membrane current and its application to conduction and excitation in nerve. *Journal of Physiology* 117(1952):500.

*Hodgkin, A. L., and Huxley, A. F. Movement of sodium and potassium ions during nervous activity. *Cold Spring Harbor Symposia on Quantitative Biology*. Vol. XVII(1952):43–52. A brief, simplified account of the results of the voltage-clamp experiments on the squid giant axon.

Hodgkin, A. L., Huxley, A. F., and Katz, B. Measurement of current voltage relations in the membrane of the giant axon of *Loligo*. *Journal of Physiology* 116(1952):424.

*Keynes, R. D. The nerve impulse and the squid. *Scientific American*, Dec. 1958. Also available as Offprint 58.

*Keynes, R. D. Ion channels in nerve cell membrane. *Scientific American*, March 1979. Also available as Offprint 1423.

*Morell, P., and Norton, W. T. Myelin. *Scientific American*, May 1980. Also available as Offprint 1469.

*Stevens, C. F. The Neuron. *Scientific American*, Sept. 1979. Also available as Offprint 1437.

Tasaki, I. Conduction of impulses in the myelinated nerve fiber. *Cold Spring Harbor Symposia on Quantitative Biology*, Vol. XVII(1952): 37–41.

Synaptic transmission

*Axelrod, J. Neurotransmitters. *Scientific American*, June 1974. Also available as Offprint 1297.

Ceccarelli, B., and Hurlbut, W. P. Vesicle hypothesis of the release of acetylcholine. *Physiological Reviews* 60(1980):396.

Coombs, J. S., Eccles, J. C., and Fatt, P. Excitatory synaptic action in motoneurones. *Journal of Physiology* 130(1955):374.

Del Castillo, J., and Katz, B. Quantal components of the end-plate potential. *Journal of Physiology* 124(1954):560.

*De Robertis, E. Ultrastructure and cytochemistry of the synaptic region. *Science* 156(1967):907.

*Eccles, J. C. The synapse. *Scientific American*, Jan. 1965. Also available as Offprint 1001.

Fatt, P., and Katz, B. 1952. Spontaneous subthreshold activity at motor nerve endings. *Journal of Physiology* 117(1952):109.

*Katz, B. How cells communicate. *Scientific American*, Sept. 1961. Also available as Offprint 98.

Katz, B. Quantal mechanism of neural transmitter release. *Science* 173(1971):123.

Katz, B., and Miledi, R. The timing of calcium action during neuromuscular transmission. *Journal of Physiology* 189(1967):535.

*Lester, H. A. The response to acetylcholine. *Scientific American*, Feb. 1977. Also available as Offprint 1352.

Matthews, G., and Wickelgren, W. O. Glycine, gaba and synaptic inhibition of reticulospinal neurones of lamprey. *Journal of Physiology* 293(1979):393.

Matthews, G., and Wickelgren, W. O. Glutamate and synaptic excitation of reticulospinal neurones of lamprey. *Journal of Physiology* 293(1979):417.

Takeuchi, A., and Takeuchi, N. On the permeability of the end-plate membrane during action of the transmitter. *Journal of Physiology* 154(1960):52.

Skeletal muscle

Ashley, C. C., and Ridgeway, E. B. Simultaneous recording of membrane potential, calcium transient and tension in single muscle fibres. *Nature* 219(1968):1168.

Bourne, G. H. *The Structure and Function of Muscle.* 2d ed. New York: Academic Press, 1972 (Vol. I), 1973 (Vols. II and III), 1974 (Vol. IV).

Buchtal, F., and Schmalbruch, H. Motor unit of mammalian muscle. *Physiological Reviews* 60(1980):90.

Caputo, C. Excitation and contraction processes in muscle. *Annual Review of Biophysics and Bioengineering* 7(1978):63.

*Cohen, C. The protein switch of muscle contraction. *Scientific American*, Nov. 1975. Also available as Offprint 1329.

Endo, M. Calcium release from the sarcoplasmic reticulum. *Physiological Reviews* 57(1977):71.

Freund, H.-J. Motor unit and muscle activity in voluntary motor control. *Physiological Reviews* 63(1983):387.

*Hoyle, G. How is muscle turned on and off? *Scientific American*, April 1970. Also available as Offprint 1175.

Hoyle, G. *Muscles and Their Neural Control.* New York: Wiley, 1983.

*Huxley, H. E. The mechanism of muscular contraction. *Scientific American*, Dec. 1965. Also available as Offprint 1026.

Huxley, H. E. The structural basis of muscular contraction. *Proceedings of the Royal Society of London, Series B*, 178(1971):131.

Huxley, H. E. Muscular contraction and cell motility. *Nature* 243(1973):445.

*Murray, J. M., and Weber, A. The co-operative action of muscle proteins. *Scientific American*, Feb. 1974. Also available as Offprint 1290.

Peachey, L. D. The sarcoplasmic reticulum and transverse tubules of the frog's sartorius. *Journal of Cell Biology* 25(1965):209.

*Porter, K. R., and Franzini-Armstrong, C. The sarcoplasmic reticulum. *Scientific American*, March 1965. Also available as Offprint 1007.

Sheterline, P. *Mechanisms of Cell Motility: Molecular Aspects of Contractility*. New York: Academic Press, 1983.

Stracher, A. *Muscle and Nonmuscle Motility*. Vol. I. New York: Academic Press, 1983.

Heart

*Adolph, E. F. The heart's pacemaker. *Scientific American*, March 1967. Also available as Offprint 1067.

*Berne, R. M., and Levy, M. N. *Physiology*. St. Louis: C. V. Mosby, 1983. Chapters on cardiovascular system.

Brown, H. F. Electrophysiology of the sinoatrial node. *Physiological Reviews* 62(1982):505.

Carpenter, D. O. *Cellular Pacemakers*. Vol. I. New York: Wiley, 1982.

Fozzard, H. A. Heart: Excitation-contraction coupling. *Annual Review of Physiology* 39(1977):201.

Noble, D. *The Initiation of the Heartbeat*. 2d ed. New York: Oxford University Press, 1979.

Vasselle, M. Electrogenesis of the plateau and pacemaker potential. *Annual Review of Physiology* 41(1979):425.